REVISING REALITY

A BIBLICAL LOOK INTO THE COSMOS
Volume One

How Dramatic New Discoveries in Biblical Texts
and Breakthroughs in Modern Physics
are Transforming Our View of the Cosmos

Anthony **PATCH** Josh **PECK** Gonzo **SHIMURA** S. Douglas **WOODWARD**

WITH AN ASTONISHING FOREWORD BY STAN DEYO

Psalm 8

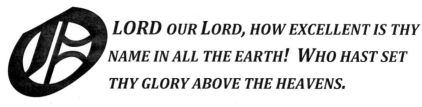**LORD OUR LORD, HOW EXCELLENT IS THY NAME IN ALL THE EARTH! WHO HAST SET THY GLORY ABOVE THE HEAVENS.**

OUT OF THE MOUTH OF BABES AND SUCKLINGS HAST THOU ORDAINED STRENGTH BECAUSE OF THINE ENEMIES, THAT THOU MIGHTEST STILL THE ENEMY AND THE AVENGER.

WHEN I CONSIDER THY HEAVENS, THE WORK OF THY FINGERS, THE MOON AND THE STARS, WHICH THOU HAST ORDAINED;

WHAT IS MAN, THAT THOU ART MINDFUL OF HIM? AND THE SON OF MAN, THAT THOU VISITEST HIM?

FOR THOU HAST MADE HIM A LITTLE LOWER THAN THE ANGELS, AND HAST CROWNED HIM WITH GLORY AND HONOUR.

THOU MADEST HIM TO HAVE DOMINION OVER THE WORKS OF THY HANDS; THOU HAST PUT ALL THINGS UNDER HIS FEET:

ALL SHEEP AND OXEN, YEA, AND THE BEASTS OF THE FIELD;

THE FOWL OF THE AIR, AND THE FISH OF THE SEA, AND WHATSOEVER PASSETH THROUGH THE PATHS OF THE SEAS.

O LORD OUR LORD, HOW EXCELLENT IS THY NAME IN ALL THE EARTH!

For Thee O LORD, and Thy vision for humanity soon to be glorified as your eternal sons and daughters, this book is dedicated

REVISING REALITY

A BIBLICAL LOOK INTO THE COSMOS
Volume One

*HOW DRAMATIC NEW DISCOVERIES
FROM BIBLICAL TEXTS AND BREAKTHROUGHS
IN MODERN PHYSICS ARE TRANSFORMING
OUR VIEW OF THE COSMOS*

Anthony Patch

Josh Peck

Gonzo Shimura

S. Douglas Woodward

Foreword by Stan Deyo

Faith Happens

Oklahoma City

REVISING REALITY:

A Biblical Look into the Cosmos

How Dramatic New Discoveries from Biblical Texts
and Breakthroughs in Modern Physics
Are Transforming Our View of the Cosmos

Copyright © Faith Happens 2016, Oklahoma City, Oklahoma

Contributing authors, alphabetically: Anthony Patch, Josh Peck, Gonzo Shimura, and S. Douglas Woodward.

Edited by S. Douglas Woodward.

Foreword by Stan Deyo.

Quotations from the Bible are taken from the King James Version unless otherwise noted.

Illustrations are taken from the Wikipedia Commons unless otherwise noted. Licensing information is defined for their usage if non-standard.

Information about this book and ordering multiple copies can be obtained by emailing info@faith-happens.com.

A very special thanks to Kim Hanke, Gary Huffman, and Audrey Robison Vanderkley for their efforts in proofing the manuscript.

V.2.0 1 October 2016

ISBN-13: 978-1537541266

ISBN-10: 1537541269

Printed by Create Space

Cover design by Gonzo Shimura

Table of Contents

TABLE OF FIGURES VIII
EDITOR'S OVERVIEW: KEY MESSAGES IN THIS BOOK XI
FOREWORD BY STAN DEYO 3
INTRODUCTION: HOW ANCIENT TEXTS AND THE NEW PHYSICS REVISE OUR VIEW OF REALITY 13
    WHY COSMOLOGY AND MEANING ARE LINKED 13
    HOW CAN WE JUSTIFY HUMANITY MATTERS AT ALL? 15
    THE COSMOS BEGINS BY GOD *CREATING* IT & HUMANITY 18
    WHY THIS BOOK WAS WRITTEN 19
    NO MATTER HOW OLD YOU BELIEVE THE EARTH IS 20
    WHY IS THERE EVIL AND DEATH IN THE WORLD? 22
    SOPHIA AND LOGOS – THE MIND OF GOD 24
    CHRISTIANITY'S LOGOS IS NOT PHILO'S LOGOS 27
    EVIL IS MORE THAN THE DARK SIDE OF THE FORCE 30
    NEW PARTICLES AND DIMENSIONS TO THE CREATION 34
    TECHNOLOGY AND THE DESTINY OF HUMANITY 35
HOW THE NEPHILIM CHANGED THE COURSE OF OUR COSMOS . 39
    FAR OUT DUDE 39
    SONS OF GOD 42
    GIANTS 45
    A COSMIC CONNECTION 48

- FALLEN ANGELS 51
- OUR HEAVENLY HOME – AN IMMORTAL HOUSE OF FLESH 54
- NOAH 59
- THE SECOND INCURSION – FIVE DIFFERENT THEORIES 61
- NEPHILIM AND THE LAST DAYS 66
- WILL HYBRID BEINGS APPEAR AGAIN IN OUR DAY? 68
- CONCLUSION 71

DO ANGELS HAVE DNA? 75
- ANGELIC GIGANTISM 75
- SUPER-PHYSICALITY AND EXTRA DIMENSIONS 78
- INTRUDERS FROM HIGHER DIMENSIONS 80
- EXTRADIMENSIONAL GENETICS 81
- MOTIVATIONS OF EXTRADIMENSIONAL ALIENS 83
- SHAPESHIFTING 84
- THE TWO TYPES 85
- NATURAL SHAPESHIFTERS 85
- CREATED SHAPESHIFTERS 87
- CONCLUSION 89

THE DIVINE COUNCIL, LESSER ELOHIM, AND THE GOVERNMENT OF GOD 91
- AN INCONVENIENT TRUTH ABOUT HEAVEN 91
- IS THERE ONE AND ONLY ONE *ELOHIM*? 91
- MEETING THE DIVINE COUNCIL IN HEAVEN #3 93
- SONS OF GOD – MORE THAN ONE OF A KIND 98
- SONS OF GOD AND FUTILITY IN THE COSMOS 103
- NO LONGER SO KODESH – NO KIDDING 108
- THE DIVINE COUNCIL AND YOU 113

THE ELECTRIC UNIVERSE 117
- OCCULTED PHYSICS 117
- EINSTEIN AND TESLA: DISTINCTIVE VIEWS OF THE COSMOS 120
- ELECTRICITY NOT GRAVITY: THE COSMIC PRINCIPLE 122
- TESLA TO THE RESCUE 124
- THE PARADIGM SHIFTS 126
- THE GOLDEN AGE OF KRONOS – AN OCCULT MYTH 128
- ANCIENT PHYSICS ON THE WALLS OF CAVES 133
- CONCLUSION 139

THE GODS OF CERN AWAKEN 141
   THE FATE OF THE WHOLE WIDE WORLD HELD IN OUR HANDS. 141
   JUST HOW STRANGE IS A STRANGELET? 143
   BREAKING UP IS SO HARD TO DO 146
   THE GODS AROUND (AND UNDER) THE MACHINE 148
   THE HIDDEN AGENDA BEHIND CERN'S COLLISIONS 150
   A DNA MAKEOVER MASQUERADING BEHIND THE LHC 152
   THE OPEN DOOR POLICY AT CERN 156
   "POWER-UPS" OF MUCH GREATER MAGNITUDES 160
   CERN & THE FICTIONAL STARGATE: ARE THEY ALIKE? 162
   BEAM US DOWN SCOTTY! 166
   FROM THE RINGS OF CERN TO THE RINGS OF SATURN 171
   CONCLUSION 175

WAS EINSTEIN WRONG? THE EMERGING COSMOLOGY—PHYSICS, ALCHEMY, AND THE NEW REALITY 179
   SCIENTIFIC REGARD FOR THE SUPERNATURAL 179
   THE NEW REALITY 183
   THE STANDARD THEORY PERSEVERES 186
   IS IT TIME FOR THE AETHER - AGAIN? 189
   DAVID HUDSON'S GOLD STRIKE 198
   THE THIRD REICH AND ITS TAKE ON THE NEW PHYSICS 202
   NORMAL VS. PARANORMAL: WHY THE DISTINCTION? 208
   CONCLUSION: POWERS BEYOND THE SUPERNATURAL 209

MAN-MACHINE HYBRID: THE NETWORKED CHIMERA AWAKENS 213
   INTRODUCTION 213
   THE NEW AGE OF THE INDUSTRIAL REVOLUTION 216
   THE CHIMERA COLOSSUS COMES TO LIFE 220
   THE METAPHYSICS OF THE TECHNIUM 223
   PROPHETIC REFLECTIONS ON THE TECHNIUM BEAST 228
   EXAMPLES OF CONFLUENCE – MAN AND MACHINE 230
   CONCLUSION 232

THE TOWERS OF THE TECHNIUM 235
   THE JASHER TECHNOLOGY 238
   AN ASSAULT UPON YAHWEH & HIS HEAVENLY HOST 242
   HEAVENLY HOSTS AND THE BABEL CONNECTION 244
   INTO MODERNITY 247

PARTICLE ACCELERATORS 249
DESERTRON 252
BRIDGES TO META-REALITY: CONNECTING THE DOTS 254
WHAT'S A SOUL TO DO IF BLACK HOLES HOUND US? 256
CONCLUSION 261

ADVANCED CIVILIZATIONS AND THE REAL STAR WARS? 263
WHEN SCIENTISTS COULD BELIEVE 263
THE SECRET SPACE PROGRAM 266
THE TRIGGERING EVENT 268
A VAST INSTITUTION – INVISIBLE & UNGOVERNABLE 271
THE COSMIC WAR—A REAL STAR WAR? 272
ION CANONS IN THE COSMIC WAR 276
WHEN DID THIS COSMIC WAR HAPPEN? 280
CONCLUSION: THE COSMIC WAR & ITS IMPLICATIONS 285

CONCLUSION: THE FUTURE QUANTUM WORLD 291
INTRODUCTION 291
PARALLEL AND EXTRA DIMENSION DETECTION 293
QUANTUM COMPUTING 296
QUANTUM TELEPORTATION 299
VACUUM (ZERO-POINT) ENERGY 300
CONCLUSION 304

ABOUT THE AUTHORS.. 305

TABLE OF FIGURES

Figure 1 - DR. EDWARD TELLER 3
Figure 2 - RUSSIAN STAMP COMMEMORATES SAKHAROV'S NOBEL PRIZE 5
Figure 3 - CAPTAIN SIR JOHN P. WILLIAMS. C.M.G., O.B.E. 7
Figure 4 – DEYO AND THE DISCOVERY OF THE GARDEN OF EDEN 9
Figure 5 – STAN DEYO, AT THE CROSSROADS OF SCIENCE AND THE BIBLE 10
Figure 6 – PRO*PHETIC PERILS* BY HOLLY DRENNAN DEYO 11
Figure 7 – A PICTURE OF DEEP SPACE FROM THE HUBBLE TELESCOPE - GALAXIES GALORE 13

Figure 8 – ASTRONOMER CARL SAGAN 14
Figure 9 - THE GOD DELUSION 16
Figure 10 - FRIEDRICH NIETZSCHE.. 17
Figure 11 - DISCOVERY INSTITUTE, ADVOCATES FOR INTELLIGENT
 DESIGN.. 21
Figure 12 - THE GNOSTIC COSMOS F 25
Figure 13 - AN ALIEN ENGINEERS HUMANITY IN THE MOVIE
 PROMETHEUS 40
Figure 14 – THE FALL OF THE REBEL ANGELS BY HIERONYMUS
 BOSCH 45
Figure 15 - THE APOCRYPHAL BOOK OF ENOCH 49
Figure 16 - GATE OF FALLEN ANGELS, MOUNT HERMON.................... 53
Figure 17 - LOT AND HIS DAUGHTERS BY LUCAS VAN LEYDEN 60
Figure 18 - *OUR POST-HUMAN FUTURE,* FRANCIS FUKUYAMA................ 69
Figure 19 – FLATLAND BY EDWIN ABBOTT, COPY OF FIRST EDITION 77
Figure 20 - A SINGLE CELL ZYGOTE, GAMETES JOIN 81
Figure 21 – BY LUCAS CRANACH THE ELDER (LANDESMUSEUM)..... 83
Figure 22 - BIZARRO WORLD SUPERMAN ADVENTURE COMICS 90
Figure 23 - FRESCO OF THE COUNCIL AT NICAEA IN THE SISTINE
 SALON 92
Figure 24 – THE DEATH OF KING AHAB... 96
Figure 25 - SONS OF GOD – DAUGHTERS OF MEN 105
Figure 26 - SHIVA, THE GOD OF DESTRUCTION AT CERN................ 120
Figure 27 - BY FASTFISSION - FROM EDWARD GRANT, "CELESTIAL
 ORBS IN THE LATIN MIDDLE AGES"... 123
Figure 28 – SYMBOLS OF AN ALIEN SKY, THE THUNDERBOLTS
 PROJECT..125
Figure 29 - DO WE REALLY DETECT JUST 4% OF THE UNIVERSE? 127
Figure 30 - KRONOS (SATURN) WITH SICKLE.......................................129
Figure 31- PICTOGRAPHS ON CAVE WALLS COMPARED TO PLASMA
 DISCHARGES IN THE LABORAORY... 133
Figure 32 - THE OCCULT SYMBOL OF THE BLACK SUN...................... 135
Figure 33 - THE FULLERENE OF BUCKMINSTER FULLER 138
Figure 34 - THE CERN LOGO - THE CIRCULAR AND LINEAR
 ACCELERATORS COMBINED? 145
Figure 35 – NOVELS WRITTEN BY ANTHONY PATCH............................158

Figure 36 – THE ATLAS DETECTOR SEARCHES FOR OTHER DIMENSIONS 161
Figure 37 – THE HEXAGONAL FORMATION ATOP SATURN: A SIX-SIDED CLOUD PATTERN 172
Figure 38 - BIRKELAND CURRENTS IN THE COSMOS....................174
Figure 39 - THE MORNING OF THE MAGICIANS............................180
Figure 40 – NIKOLA TESLA READING ROGER BOSCOVITCH'S BOOK, *Theoria Philosophiae Naturalis* 184
Figure 41- NIKOLAI KOZYREV...190
Figure 42 – WILIGUT'S VERSION OF THE TRUINE STONE....................205
Figure 43 – THE INSIGNIA OF THE THULE SOCIETY 207
Figure 44 – THE HIGHLY REGARDED WORK BY E.M. FORSTER, THE MACHINE STOPS...220
Figure 45 - THE COMPANY WHICH CREATED SKYNET IN THE *TERMINATOR* MOVIE SERIES. 221
Figure 46 – KEVIN KELLY, MASTER OF THE TECHNIUM 227
Figure 47 - THE BRICK WORK OF THE GREAT BATH 237
Figure 48 - A PAINTING BY PIETER BRUEGEL THE ELDER CALLED "THE TOWER OF BABEL (VIENNA)." 239
Figure 49 - VIRGIL SOLIS: "GOD'S COUNCIL" 245
Figure 50 -THE RINGS OF ACCELERATORS AS INFINITY (8)............249
Figure 51 - THE PARTICLE ACCELERATOR THAT NEVER WAS. 252
Figure 52 - THE LARGEST STONE AT BAABEK IN LEBANON 253
Figure 53 - TOLKIEN'S *TWO TOWERS* 262
Figure 54 – SAGAN AND SHKLOVSKII, 1966 263
Figure 55 –JOSEPH P. FARRELL 267
Figure 56 – THE DAY AFTER ROSWELL..269
Figure 57 – THE COSMIC WAR 274
Figure 58 –EXAMPLES OF RONGORONGO PETROGLYPHS 279
Figure 59 – EASTER ISLANDS MOAI 281
Figure 60 – THE FACE ON MARS 289
Figure 61 - CHERUBIM CHARIOTS BY JOSH PECK........................292
Figure 62 – D:WAVE– MAKER OF QUANTUM COMPUTERS 297
Figure 63 – *QUANTUM CREATION* BY JOSH PECK 298
Figure 64 - RAY KURZWEIL: THE SINGULARITY BY 2045? 303

EDITOR'S OVERVIEW:
KEY MESSAGES IN THIS BOOK

CHRISTIANS TODAY LACK WISDOM CONCERNING THE NATURE OF THE CREATION. SCIENCE HAS TAUGHT US MANY FALSE THEORIES ABOUT THE STRUCTURE OF THE UNIVERSE FROM THE smallest particle to the largest galaxy. New scientific discoveries as well as authentic insights into the ancient text of the Bible give occasion to revise our understanding in accordance to what the Lord God would have us know about Him and His creation. This book blends the efforts and insights of four researchers and writers, drawing further wisdom from the writings and teachings of many more scientists and biblical scholars. Our intent is to enable you, the reader, *to revise your understanding of reality.*

The first and foremost premise is that the Cosmos was created by the Word of God. *"In the beginning, God created the heavens and the earth"* (Genesis 1:1, ESV). *"In the beginning was the Word, and the Word was with God, and the Word was God. He was in the beginning with God. All things were made through him, and without him was not any thing made that was made."* (John 1:1-3, ESV) <u>Cosmology begins with God creating.</u>

The Bible is a book that begins and ends with Cosmology. God created an "ordered creation" according to the witness of Genesis. However, the creation appears to have been in a chaotic state, a "wasteland" formless and void when He restored it. It was deluged with water from a prior state. After its restoration, the creation was very good.

- *Genesis 1:2 - "And the earth was waste and empty, and darkness was on the face of the deep, and the Spirit of God was hovering over the face of the waters." (Darby)*

- *2 Peter 3:5-7, 10 - "For they deliberately overlook this fact, that the heavens existed long ago, and the earth was formed out of water and through water by the word of God, and that by means of these the world that then existed was deluged with water and perished. But by the same word the heavens and earth that now exist are stored up for fire, being kept until the day of judgment and destruction of the ungodly. But the day of the Lord will come like a thief, and then the heavens will pass away with a roar, and the heavenly bodies will be burned up and dissolved, and the earth and the works that are done on it will be exposed."*

Humanity will one day soon be glorified to reign with God over the Cosmos. God created humankind as men and women. In His Image He created them. He has redeemed the creation through Christ and will recreate the Cosmos restoring humanity to its proper place – in fellowship with Him and to rule and reign over the Creation with Him.

- *Genesis 1:27 - "So <u>God created man in his own image</u>, in the image of God he created him; male and female he created them." (ESV)*

- *Romans 8:19-21 - "For the creation waits with eager longing for <u>the revealing of the sons of God.</u> For the creation was subjected to futility, not willingly, but because of him who subjected it, in hope that <u>the creation itself will be set free from its bondage to corruption and obtain the freedom of the glory of the children of God.</u>"*

- *Revelation 21:1-5 – "<u>Then I saw a new heaven and a new earth, for the first heaven and the first earth had passed away, and the sea was no more</u>. And I saw the holy city, new Jerusalem, coming down out of heaven from God, prepared as a bride adorned for her husband. And I heard a loud voice from the throne saying, "Behold, the dwelling place[a] of God is with man. He will dwell with them, and they will be his people, and God himself will be with them as their God. He will wipe away every tear from their eyes, and death shall be no more, neither shall there be mourning, nor crying, nor pain anymore, for the former things have passed away." And he who was seated on the throne said, <u>"Behold, I am making all things new."</u> (ESV)*

The universe has a unifying "atmosphere" which permeates all of reality. This "Cosmos" has given birth to galaxies, stars, and planets, at the macro-end of existence while it consists of atoms and quanta: the smallest particles at the micro-end of existence, e.g., quarks, gluons. The standard academic view asserts 96% of the universe cannot be discerned by humanity through any of its instruments. We believe this is one of the principal falsehoods of academically accepted science. The "standard model" asserts that space is a vacuum. In contrast, <u>we believe space is filled was ionized gas known as *plasma*.</u> The plasma serves as a medium throughout the universe to propagate electromagnetic waves between heavenly bodies, just as air propagates sound waves within the atmosphere of earth. This substance is known as the "aether" (the Greek word for atmosphere). Science has debated the presence of the aether for centuries. The Bible seems to imply that the aether exists throughout the heavens. Many scientists today that challenge the standard model assert this aether has been proven to exist in numerous way.

Secularism and atheism oppose what the Bible teaches about how the Cosmos began and how it will end. "Scientism" has persuaded us all into believing the universe began with a Big Bang and expands endlessly until it eventually dies a cold death of entropy billions of years from now. Paradoxically, despite this dismal fate, science asserts the universe is guided by a non-sentient evolutionary principle that thrusts us toward progress. Supposedly, life inevitably exists and evolves everywhere throughout the universe. Evolution cannot be detected by instruments or proven in the laboratory. Its evidence in geology and biology is inconclusive. However, the standard academic viewpoint asserts that evolution is a proven fact. Thus, Scientism demonstrates that it is built upon unprovable axioms and thus, draws conclusions about the nature of reality that become a matter of faith more than science.

The Cosmos, from a secular atheistic perspective, does not vindicate any claims to "right or wrong". For morality to exist with "ultimate consequences" for doing good or doing evil, there must be a God to enforce right values. Modernity generally rejects any notion of a personal God (Theism) and therefore, assumes that morality and the importance of following conception of "right and wrong" are baseless. The only way to ground moral action is based upon humanity's "will to power" (i.e., willing it so). According to infamous German philosopher, Friedrich Nietzsche, <u>since God does not exist, morality does not exist.</u> Thus, at its core, the universe is amoral. Humanity's sentiments toward right and wrong are not grounded in a Divine Judge because no such judge exists. From the vantage point of atheism, only the *Übermensch* can resolve this dilemma. Man must choose what's right and wrong, period.

Our reality is fluid (unstable) and threatened by natural mechanisms "built into" the universe. According to Einsteinian physics, the speed of light is a constant that cannot be exceeded; gravity is the dominate force in the universe; existence is haunted by black holes and is comprised mostly of dark matter and dark energy. Time and space are interwoven. Energy consists of mass multiplied by the speed of light, squared ($E=mc^2$). Objects with great mass can bend space and time. What seems "solid and sure" is actually relative to the mass of an object and its speed traveling through the Cosmos. Furthermore, scientists now assert we cannot really be sure our experienced existence will remain constant and "settled" (our past, present or future) as parallel universes may exist promising potentially different outcomes and altered histories.

Humankind's religions are built principally around three diabolical ideas. One: the creation is flawed because matter is flawed. Evil results from the flaws embedded within our reality. Whoever made the creation did not have the will or ability to make it good. Two: humanity is a creature able to achieve godhood and/or reach immortality after this life. We can overcome the limitations of the flawed creation by recognizing a spiritual reality that supposedly suffuses the entire universe. It is called by many names. In our day, the notion was once entitled "cosmic consciousness". This divinity can be accomplished through following particular religion practices, based upon acquiring special knowledge (gnosis) or conducting rituals (prayer, conduct of good deeds, obeying religious laws). We must accept the axiom that spirit supersedes matter. According to most religions, matter may be an illusion. Reality itself may be an illusion. Even some conservative Christian theologians suggest reality may only be a *hologram* – a projection of the mind of God. Three: human religion falsely assumes that power exists that enables us to achieve complete enlightenment. This power is sometimes called "the being of light" or Lucifer. It comprises a force or principle in the universe which expands our capacity to understand and to influence existence, and wants to guide humanity to fulfillment. Even a "non-religion" like Freemasonry, asserts that the initiated can learn cosmic secrets and the god Lucifer will instruct them.

REVISING REALITY

Philosophy of religion posits that good and evil are either impersonal forces that exist in the universe or complete illusions that do not exist at all. In opposition, Judeo-Christian teaching asserts that they are *personages* **not** forces: Jehovah is the first such person who exists prior to everything else and is the Elohim of elohim. He created every other personage in the entire creation: whether seen or unseen by human beings. He is known to Jews as Yahweh, but to Christians as Father and Son: His name is Yahweh or Jehovah. He is also the "Angel of God", Y'shua, or Jesus Christ. His Holy Spirit comes to dwell in us if we receive Him and welcome Him into our lives. Christians see this Spirit as fully *personal* also, the third member of a three-in-one godhead, the Trinity). Yahweh's principal adversary is Satan (which means adversary) who turned against Him at some point in the past. He stands opposed to God, but he is not the opposite of God. His name, Satan, means *adversary*. His rebellion was at least thousands of years ago in terms of earthly chronology – or perhaps even further back, in eons past, when Satan first sought to usurp Jehovah's rule over all creation.

God created both heavenly and earthly beings, choosing to rule with and through them. He created a *Heavenly Host* or *Divine Council* (see Psalm 82 and 89), as well as angels that now exist in both righteous and unrighteous states. He ultimately created humanity. Humanity also rebelled against Yahweh. As a result, it now exists infected with an inclination to sin against other humans, the creatures of this world, and God Himself. Nevertheless, humanity has been redeemed by a unique God-man, Jesus Christ. Through Y'shua, God has made provision for humanity to rise above this corruption. This redemption, however, does not apply to *the fallen angels*. All truly redeemed human beings reflect <u>in part</u> the promise of the coming new world and a perfected humanity. Until the resurrection at the return of the God-man – Y'shua, at the end of this age – humanity will continue to experience death and the consequences of sin, especially if and when we lapse habitually into sin. God's provision, however, does promise freedom from this sin inclination *if we live in the Spirit drawing power from God Himself living in and through us.*

The creation was corrupted yet again upon the disobedience of the first humans, Adam and Eve. According to the Bible, the tempter was Satan, the *nachash*, the Hebrew word for *snake*. Later, we are told that a group of angels fell from heaven (Genesis 6). These angels came to Earth and had offspring with the "daughters of Adam", called the *Nephilim*. A non-biblical text, *The Book of Enoch*, expands on Genesis' very brief account. It is quoted in the Bible, but cannot be considered authoritative. Nevertheless, it is an ancient text worthy of study. According to its account, 200 angels descended upon Mount Herman (today's Golan Heights) and inbred with human women creating hybrid offspring. This event was the catalyst for the mythology of virtually all ancient peoples (Sumerians, Babylonians, Egyptians, Greeks, Romans, and Nordic peoples). Many evangelical scholars who have studied this account in depth, in particular Dr. Michael S. Heiser, explains that whether intentional or not, the impact of this angelic incursion into the human race was used by Satan to defeat Yahweh's plan to redeem humanity setting the creation in order once more. <u>This event was non-trivial.</u> The Christian notion of salvation through the blood of Jesus Christ builds upon rightly interpreting this event. Without this appreciation, Old Testament accounts of giants and the slaughter of the peoples of Canaan by the Hebrews as led by Moses and Joshua, lead to unbiblical interpretations of the nature of God. Additionally, it should be pointed out that the Flood of Noah was (at minimum) God's interim plan to eradicate hybrid humanity created by these fallen angels and animal chimera they produced. Through God's plan revealed to Noah, only selected animals (pure ones) accompanying Noah and his family on an ark would preserve life and perpetuate these species on this earth.

Nevertheless, after the flood, the Nephilim reappear. Why did these hybrid beings return once more? There are various explanations for their reappearance. In *Revising Reality*, the authors discuss the possibilities, examining five different theories. They also consider the special abilities of angels, their multidimensionality, whether they might have DNA, and whether Nephilim will appear again in these last days, effectively mirroring "the days of Noah", for it has prophetic import.

Christians today live without awareness that alternative physics reopens the possibility of the supernatural. Although the Bible does not distinguish between the "natural and the supernatural", humanity does. Since the eighteenth century, the period known as "The Enlightenment", "standard science" has rejected the supernatural events such as the Bible describes. *All effects have natural causes.* <u>There are laws that govern the workings of the universe and these laws can be identified and understood.</u> A supernatural event would overrule this view. Despite his deep commitment to a biblical understanding of the Cosmos, Sir Isaac Newton's understanding of physics and a much less theistic Albert Einstein's revision of Newton in the twentieth century, provide the vast majority of what science has taught us about physics including *reinforcement of an anti-supernatural bias.* However, two other scientists confronted the views of Newton and Einstein: Werner Heisenberg and Nicola Tesla. Heisenberg challenged the validity of Newton and Einstein at the quantum level (the tiniest "spaces" in the universe), while Tesla challenged Einstein concerning the dominance of gravity, the "bending" of space, the prominence of the aether (plasma), as well as proposing methods to defeat gravity through electrogravitics. Thus, Christians must now recognize that what they were taught in school may be wrong. Popular astronomers and cosmologists today continue to reinforce many falsehoods. However, orthodox science as expounded by Newton and Einstein is no longer sacrosanct. There are growing numbers of scientists that challenge the academically accepted cosmology. <u>Thus, there are substantive new insights in physics itself which call for a revised reality</u>.

Christians are unaware there is a new school of science that better reflects biblical cosmology and rejects secular/atheistic cosmology. Although scientific orthodoxy still insists upon the notions of dark matter and dark energy, black holes, and *the constant of the speed of light,* there are now an increasing number of scientists that question the validity and universality of these assertions. In particular, a new school of scientists have begun to propound *The Electric Universe.* This group of mavericks comprise the *Thunderbolts Project.* They have been rethinking the nature of the Cosmos. Their ideas acknowledge much of the so-called "standard model"; however, they depart in a number of areas based upon the work of Tesla and the notion of the fourth state of matter and energy (i.e., *plasma*). According to them, the dominant force in the universe is not gravity but electromagnetism. They point out that even Einstein could not reconcile certain "real-world" phenomena with his theories. In particular, particle entanglement, also known as "spooky action at a distance" infers that the speed of light is no longer a constant (it can be exceeded). Entangled particles "communicate" instantly no matter the distance between them. Because of these new developments, our conception of reality is in need of a vast revision. Practical applications of this quantum reality are being put into place today. Many more will come.

Whether following the lead of Einstein or of Tesla, all scientists now believe that the universe consists of multiple dimensions that we cannot see or detect. To some indefinite extent, the Cosmos is a "multiverse". There are perhaps as many as ten dimensions plus time. (Whether time in fact truly constitutes a dimension remains a matter of debate too). Because humanity lives within these three physical dimensions of height, width, and depth, it is not obvious to the vast majority of human men and women that there could be other dimensions. Nor is it apparent how we can understand them. Scientists remain certain of the multiple unseen dimensions due to mathematical equations that "make sense" of the un-sensed world. Those persons who profess a so-called "sixth sense", have a much easier time accepting the multi-dimensional hypothesis. Therefore, since so many acknowledge the multiverse, Christians must revise their reality. These vistas open the door to "supernatural" events without compromising true science.

Humanity faces new threats from scientists that also are committed to the occult or pagan views of the nature of the Cosmos. These threats arise from an elite that secretly embraces the new physics of an electromagnetic universe along with commitments to what Christians would consider satanic forces that exist beyond the three observable physical dimensions of nature. The consummate work of this group is the greatest machine ever built by humanity, the large hadron collider (LHC) at CERN near Geneva, Switzerland. Scientists at CERN admit they seek to discover the existence of all types of quanta such as the so-called "god particle", the Higgs-Boson. However, as they increase the power of the collisions there, they create dangerous particles known as "strangelets" that sink to the center of the Earth and ultimately, over time, could transform our planet into a neutron star. Even physicist Stephen Hawking contends scientists at CERN "know not what they do". Strangely, CERN constitutes a new "Tower of Babel" in several respects. 66 nations have united, actively supporting the effort. However, virtually all nations globally acknowledge the work at CERN and hope to exploit its benefits. The leadership behind CERN through its public "mouth pieces" states that their goal is to "open other dimensions" through portals created by breaking the "strong force" (the gluon) and creating a gap through which other universes or other entities can access us and we them. Thus, they hope to "reach unto heaven". Unofficially of course, CERN may be creating exotic weapons. In fact, two authors of this book propose that like Nimrod who launched a projectile into heaven (according to the *Book of Jasher* it returned to Earth tainted with blood!), the forces behind CERN may be seeking to "kill God" in an ultimate statement of liberation from and defiance against humanity's maker. As at Babel on the plain of Shinar, the effort has brought *the world together to act as one*. Like humankind working under Nimrod's rule, CERN is the focal point that seeks for those who endorse and support it to "make a name for themselves" (as was the express purpose at Babel) – to be as the demigods, the Nephilim, the offspring of angels. Today's movement known as *Transhumanism* reinforces this objective. Some say we now live in a post-human world because humanity genetically is under modification.

The authors believe the meaning of the Tower of Babel comprised a number of significant but ultimately failed attempts: an assault upon God, reopening a pathway to and from the "heavenly dimension", a statement of defiance that humanity would not allow God to destroy the human race again, and for humanity collectively to become Nephilim or demigods once more apparently in the manner of Nimrod who "began to be a mighty one in the earth" (Genesis 10:8) through a means presumably shown him by fallen angels although not disclosed in the Bible or apocryphal books like *Enoch* and *Jasher*. The goal of CERN likewise seeks to defy God, deify humankind, establish human supremacy, "create a 'stairway to heaven'" once we learn how to sustain gaps in space/time at the gluon level (the "strong force"), and create portals to other dimensions, accessing other beings. <u>This will open the abyss.</u>

Our notion of what it means to be human is on the cusp of a radical change. Genetic science as well as information technology are attempting to advance humanity, to evolve humanity to a new state, in which humanity gains immortality and directly "interfaces" to one another and to all machines through a ubiquitous (universal) "internet" transcending all current conceptions. However, the possibility exists that machines will commandeer this transformative process and overwhelm humanity. We could find ourselves slaves to the machines that we have created. Such scenarios behind movies like "The Terminator" are within decades of possibility. Indeed, the likelihood that "true artificial intelligence" will develop is at the core of the thesis advanced by inventor and intellectual Ray Kurzweil concerning the "singularity" which he has predicted will be achieved by 2045. At that point, humanity will no longer control its own destiny. It will be unable to manage advancements in technology. Instead, it is theorized that technology itself will likely control humanity. Thus, the final fate of humanity will then be in question, perhaps gravely so.

The real history of our Earth, our sun, our moon, and the entire solar system needs reexamination. Where did life originate? Has our present sun always been our sun? Is the Moon an artificial satellite? Before humanity lived on Earth, did it live on Mars? Why are their petroglyphs around the world that illustrate electromagnetic phenomenon and illustrate undeniable similarity? Was the world created less than 10,000 years ago? Did cataclysms such as Immanuel Velikovsky proposed in his controversial book, *Worlds in Collision* (1950),[1] actually take place? Were there

---

[1] "With this book Immanuel Velikovsky first presented the revolutionary results of his 10-year-long interdisciplinary research to the public, founded modern catastrophism

ancient civilizations with different, perhaps more advanced technologies than what we understand today? What is the empirical evidence to support such radical revisions of reality? These issues will be explored in the Volume Two of *Revising Reality: A Biblical Look into the Cosmos*.

S. Douglas Woodward

Oklahoma City
September, 2016

– based on eyewitness reports by our ancestors – shook the doctrine of uniformity of geology as well as Darwin's theory of evolution, put our view of the history of our solar system, of the Earth and of humanity on a completely new basis – a nd caused an uproar that is still going on today." (From Amazon's summary of the book).

REVISING REALITY

"For the scientist who has lived by his faith in the power of reason, the story ends like a bad dream. He has scaled the mountain of ignorance; he is about to conquer the highest peak; as he pulls himself over the final rock, he is greeted by a band of theologians who have been sitting there for centuries."

Robert Jastrow

The Enchanted Loom: Mind in the Universe

REVISING REALITY

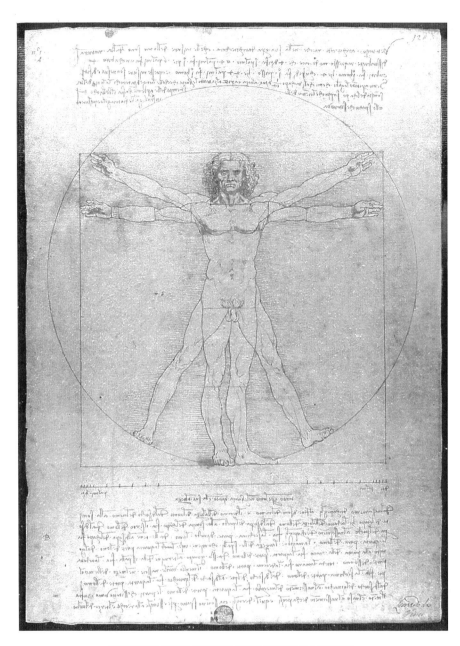

VITRUVIAN MAN BY LEONARDO DA VINCI
Photo by Luc Viatour / www.Lucnix.be

FOREWORD
BY STAN DEYO

The pursuit of truth about our existence is a noble, difficult, and perhaps dangerous pursuit – much like running a marathon and enduring until crossing the finish line. Standing on the shoulders of men like Da Vinci, Newton, Faraday and Einstein we can peer much further into the depths of our universe and perhaps even into the reality of other universes concentric to our own.

The four authors of *Revising Reality: A Biblical Look into the Cosmos*, are part of a growing cadre of clarion-wielding followers of Y'shua Ben Elohim. They, like myself and my wife, Holly (along with a finite number of others) probe the nuances of the Biblical prophecies given for our time – a time at the end of this age.

I am older than these four authors and probed these things as far back as the late 1960s. In retrospect, my path looks like an adventure in *Lord of the Rings* or perhaps one from an *Indiana Jones'* novel. No matter how one couches it, I am certain the path I followed was meant to prepare me to share what I've learned in the last half-century or so with such new "watchmen" of this century. I'm pleased to hand down these insights.

Figure 1 - DR. EDWARD TELLER

As some readers may know, in 1971 I was recruited into a multinational and above-top-secret project under the direction of men like Dr. Edward Teller

(1908-2003) from the United States and father of the hydrogen bomb, Captain Sir John P. Williams a famous seaman from Australia (1896-1989), and Dr. Andrei Sakharov (1921-1989), a Russian nuclear physicist.[1] I was told that these men wanted me to join one of their research teams in Australia to continue my investigation *into antigravity and advanced propulsion techniques* with other like-minded physicists and engineers. They flew me to Australia under the oversight of Sir John in Melbourne (mentioned above). Once there, I was introduced to Dr. Tom Keeble, Director of the Australian Aeronautical Research Labs at Fishermen's Bend in Melbourne. We met in his office. Also attending the meeting were a young atmospheric physicist and another interesting older man, typical of the exceptional people there.

---

[1] **Dr. Edward Teller** made many contributions to nuclear science. He "was an early member of the Manhattan Project, charged with developing the first atomic bomb. During this time, he made a serious push to develop the first fusion-based weapons as well, but these were deferred until after World War II. After his controversial testimony in the security clearance hearing of his former superior at Los Alamos Laboratory, J. Robert Oppenheimer, Teller was ostracized by much of the scientific community. He continued to find support from the U.S. government and military research establishment, particularly for his advocacy for nuclear energy development, a strong nuclear arsenal, and a vigorous nuclear testing program. He was a co-founder of Lawrence Livermore National Laboratory (LLNL), and was both its director and associate director for many years." Retrieved from https://en.wikipedia.org/wiki/Edward_Teller.

Sir John P. Williams was a Welshman who immigrated to Australia. He was Chairman of the Australian Coastal Shipping Commission from 1956 to 1971. He was also a member in good standing of the prestigious "Melbourne Club" which is thought to be one of the three Melbourne clubs which supplied local membership for the Australian branch of the "Round Table Society" [aka Round Table movement]. What is not widely known is that he was also a member of a multinational group of scientists, engineers and industrialists coordinated by Drs. Edward Teller and Andrei Sakharov." From citation by Deyo, http://www.standeyo.com/stans.files/Cosmic_Conspiracy_items/Sir_John_Williams.html. Additionally, "Historian Carroll Quigley claimed that the Round Table Groups were connected to a secret society, which South African diamond baron Cecil Rhodes is believed to have set up with similar goals. Rhodes was believed by some to have formed this secret society in his lifetime. This secret society is supposed to have been named the *Society of the Elect*". See https://en.wikipedia.org/wiki/Round_Table_movement.

Dr. Andrei Sakharov was "a Russian nuclear physicist, Soviet dissident, an activist for disarmament, peace and human rights. He became renowned as the designer of the Soviet Union's Third Idea, a codename for Soviet development of thermonuclear weapons. Sakharov later became an advocate of civil liberties and civil reforms in the Soviet Union, for which he faced state persecution; these efforts earned him the Nobel Peace Prize in 1975. The Sakharov Prize, which is awarded annually by the European Parliament for people and organizations dedicated to human rights and freedoms, is named in his honor." Retrieved from https://en.wikipedia.org/wiki/Andrei_Sakharov.

At that time, I was advised of a large video and photo UFO collection held by the RAAF. This began a long and challenging journey to understand the purpose of the project and eventually, the hidden government of this planet.

The Allies had begun the project many decades earlier as WWII was coming to a close. The "nobility" of the Earth's most rich and powerful men held numerous private meetings to decide how to unify the nations of the world and to limit the use of nuclear energy as a weapon. Some might label these people the "Illuminati" or the "Globalists" or some other variant. Believing it was their destiny as well as responsibility to preserve and protect mankind, they considered ways to convince the world's population (not so much the leadership – just the common people) to pledge their allegiance to a New World Order. This "Order" wanted to establish a common religion, economy, law, and language across the planet to minimize the possibility of a global nuclear war.

Figure 2 - RUSSIAN STAMP COMMEMORATES SAKHAROV'S NOBEL PRIZE

They saw that a democratic process like the old League of Nations or the current United Nations would not work as there would be no consensus on the most important issues facing humanity or the process of addressing imminent global threats on a timely basis. In effect, due to extended deliberations so typical in such institutions, organizations like the U.N. couldn't move fast enough to resolve urgent issues and imminent threats.

These men also recognized that forming a dictatorial government would not be stable if the subjects did not support it. The French resistance proved that convincingly to Hitler and the Nazis. They had but one alternative remaining. It came to be known as "Alternative Three." They would need to coerce the people into accepting a global government under an especially sanctioned king appointed and supported by an "alien" race with no political, religious or race agendas. Orson Welles' radio broadcast presenting a fake alien invasion proved that *people worldwide would pay attention and would fear the arrival of an alien race with advanced technology that made our best weapons a joke.* This was the pretense.

This idea had been discussed in passing beforehand, but not seriously until post-WWII, once the atom bomb made the issue imperative. I'm not sure when exactly, but sometime between 1943 and 1950 "other-worldly" aliens (appearing as *extraterrestrials*) made contact with our leaders. The agreements reached then gave humankind new and wonderful technologies (including antigravity propulsion) *in return for mankind's building secret underground and undersea facilities which were to be used* for the "aliens" to manufacture items needed to remain here.

Although I never went to the Antarctic base, I was told we had a facility there. As legends allege, it sounded like it could have been at Neuschwabenland (Antarctica) with former Nazis working there after WWII. It's also possible that the Germans themselves had made contact before the Allies did – and that's why they were ahead of the U.S. in the development of *electrogravitic* craft (a view held by many, since "flying saucers" were undisputedly under development in the 1940s). I know our project did underground testing of smaller disks after the war using old Nazi-excavated tunnels in Germany. The disks were mounted sideways on rail cars and then accelerated down a tunnel on old rail lines (tracks).

Today, I seldom think about those days as the project was discontinued when our leaders lost control of the joint facilities in the late 1970s. There were a few of us, those new to the project, that were cut loose from the mission. Our records having been destroyed, we were encouraged to go underground as no one knew what would happen after we lost the war with the "aliens". Obviously, I survived as did only a few others. Today, most are still afraid to disclose what they did then or what they know now.

As for those "aliens", it took me a few years to figure out who they really were. After accepting Y'shua Ben Elohim as my Messiah, I realized the "aliens" so-called, were what the Bible refers to as "fallen angels". They came from a *parallel* (perhaps better termed *concentric*) universe, which surrounds us, but exists at higher energy densities. This creates a barrier or a wall between the different realities. We can't see it, although we can pass through it at some altitude above the Earth. Regardless... it's there.

The ancient Sumerian, Babylonian, Egyptian and Hebrew records consistently mention how *gods* (super-beings) appeared in the sky with fire and heat – then they descended to Earth. We recall that Nimrod even tried to build a tower allowing him to reach the barrier or *penetrate the veil* as the Bible calls it (see Genesis 11). Before his task was completed as planned, Yahweh destroyed the Tower of Babel and immediately afterward, the people of Earth scattered. The Bible says their languages were confused to prevent them from gaining access to this realm again.

As far as I can determine, these "fallen ones" were a group of *super beings*, who rebelled against our Creator within His domain (a parallel existence or universe). They were forced out of His reality and down to Earth, to await final judgment for their foiled coup. To me, it seems likely that the deal between the Globalists and the "aliens" was struck be-

Figure 3 - CAPTAIN SIR JOHN P. WILLIAMS. C.M.G., O.B.E.

cause the rebels were shoved through the barrier with limited means to use their technical know-how to pursue their ends. I believe the Fallen Ones sought to build weapons on Earth to prepare an assault against the Creator (Y'shua Ben Elohim and His mighty army) when He comes in judgment (Revelation 19:11-21). Therefore, the "aliens" made agreements with the

Globalists. Humans would assist in building infrastructure (the underground and undersea bases alluded to earlier). In return, the Fallen Ones would transfer technology to humanity. These facilities were used for weapons development, employing technology similar to that which they had on the other side of the "dimensional" barrier. Once they reached a level of infrastructure deemed to be self-sustaining (housed in these secret facilities), they kicked the Globalists and our project teams out.

I remember a lunch with Sir John and two other members of the project. Sir John expressed his concern that the "aliens" were moving in on the humans and sought to force us out of the bases we helped them build. That was 1972. 44 years later, I have no doubt the stage has been set for the coming battle between Satan and his Earth-based army of these Fallen Ones, assembled against the coming army of the Messiah which will be launched from the other side of the "divide" – the realm of the Almighty.

It's important that information about these things and other much more current events (such as what is discussed in this book regarding Transhumanism and the troubling activities at CERN in Geneva) are shared with those who want to know… indeed, who really *need* to know. I realize my testimony will be met with disbelief if not anger by some well-meaning but close-minded people. The prophet Daniel even stated that such a day would come at the end of this age. It is most interesting that 300 years ago, Sir Isaac Newton mentioned Daniel's warning in his letters and papers. Daniel predicted a body of people would rise up to tell the truth of the times in "those days":

> *And they that understand among the people shall instruct many: yet they shall fall by the sword, and by flame, by captivity, and by spoil, many days.*
>
> *Now when they shall fall, they shall be assisted with a little help: but many shall cleave to them with flatteries.*
>
> *And some of them of understanding shall fall, to try them, and to purge, and to make them white, even to the time of the end: because it is yet for a time appointed.* (Daniel 11:33-35)

Perhaps what we see through the authors of this book, jointly and individually, are members of that body of truth-tellers. Certainly, these authors aren't alone in this quest for the truth. My wife and I along with other dedicated fellow Christian authors and lecturers are aggressively revealing various facets of the truth about current events and how they appear to fulfill Biblical prophecy. We believe the Bible to be *inerrant;* moreover, some of its messages scaled until the right moment when they are to be revealed. Because of this, we are discovering things that could not have been understood

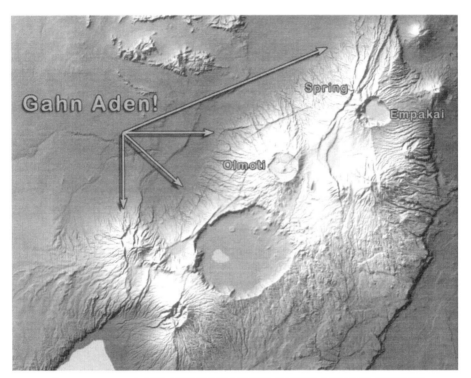

Figure 4 – DEYO AND THE DISCOVERY OF THE GARDEN OF EDEN

until today. By digging into the ancient languages of biblical revelation, supplemented by modern technology, we are expanding our understanding of many seemingly "odd" events in the ancient world – events that enliven our awareness. Allow me to enumerate several such discoveries I've made recently that illustrate this fact.

First, I discovered the location of the actual Garden of Eden in this way.

Likewise, I discovered the mechanisms which our Creator employed causing the great Flood of Noah's time. In turn, this led me to discover the real location of the legendary Atlantis, which turned out to be a large island (now a peninsula) in the Middle East where I contend the fallen ones landed and divided up the Earth among themselves. It was at this same time that the fallen ones mated with Earth women to produce the giant offspring (the Nephilim) mentioned in the Genesis 6 account. It really should be no surprise that the parallel account, the last paragraph of Plato's "Atlantis passage", recounts how the *Atlanteans* grew evil and corrupt. It was then they were punished by Zeus (the "God of gods"). Doesn't this reflect what the Bible says about the people living contemporary to Noah before the Flood? It certainly seems so to me.

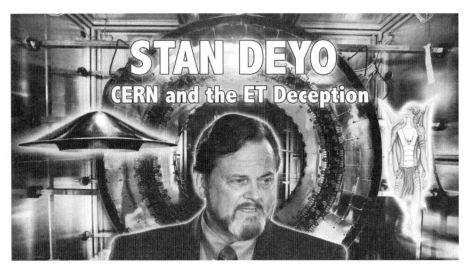

Figure 5 – STAN DEYO, AT THE CROSSROADS OF SCIENCE AND THE BIBLE

After reading *Revising Reality,* I found myself reassessing some of my own conclusions on matters such as the mind-twisting question plaguing Jewish rabbis for generations. For instance, *"How did Og and all the giants Israel fought when conquering the Promised Land survive the Flood or reappear after it had occurred?"* The Bible is not explicit on this issue. *The authors have carefully considered this matter.* Their supposition amounts to just one example of how the authors break new ground, enumerating five different ways this question can be resolved.

As the authors go on to discuss Moses and the Book of Genesis, I found myself wanting to sit them down in a quiet place so that I could share some of the evidence I found that suggests Moses compiled parts of Genesis from clay tablets personally signed by Adam, Abraham, and Noah. This is exciting. And it is the nature of theological discovery. It is how biblical knowledge grows.

And of course, these thinkers have dived into today's science – Transhumanism, genetic manipulation, the man-machine "chimera", and the implications of the activities being carried out at the Large Hadron collider. They are not timid to speculate what the meaning of these events are to humanity.

I am encouraged here to share what a rabbi privately divulged to me about the "carved" letters on the first set of tablets that God made for Moses. Apparently Moses held the tablets in the air – one in each hand – and slowly rotated them. As he did, the letters stayed in the air in front of the tablets and did not rotate with the tablets themselves! Then Moses said to them, *"This and more you have lost because of your sin!"*

Figure 6 – PRO*PHETIC PERILS* BY HOLLY DRENNAN DEYO

When he threw the tablets down the hillside in front of the elders of Israel on Mount Sinai, the tablets broke on the ground. The letters briefly danced in the air and then were gone. The fragments were smooth with no trace of carved letters. It leads me to conclude the letters were somehow projected from within the tablets like holographs. It was as if the words were alive. It was as though the tablets were a digital library of information that projected God's truth holographically. They likely could have contained much more than the Ten

Commandments. The capacity of such a tablet to preserve and present God's words would have been enormous.

My wife Holly and I are both authors. Her latest work, *Prophetic Perils – End Time Events Revealed,* is her contribution to that "body of knowledge proclaimed by truth-tellers". My first book, *"The Cosmic Conspiracy"* was my contribution to that cache of information devoted to the declaring prophetic discovery. I'm happy to expand on what I have shared before in this foreword. In fact, reading this book already has me disclosing substantive facts of my past that I have not written down before – but needed to do so. I pray this book will challenge you in the same way. Too many rest sedately in deep slumber unaware of what will soon happen despite the reality that these occurrences will be mind-numbing events certain to change everything we believe about the Cosmos. This book starts us on that path.

It's time to move: let's wake up the world!

Stan Deyo

September 2016

http://www.standeyo.com/

INTRODUCTION:
HOW ANCIENT TEXTS AND THE NEW PHYSICS REVISE OUR VIEW OF REALITY
S. Douglas Woodward

WHY COSMOLOGY AND MEANING ARE LINKED

"Cosmology: the study of the nature, origin, and evolution of the universe."
OXFORD DICTIONARY OF PHYSICS

THE MOST POPULAR SCIENTIFICALLY-THEMED PUBLIC TELEVISION PROGRAM EVER PRODUCED WAS THE 1980 PROGRAM FEATURING THE FAMOUS ASTRONOMER CARL SAGAN. THE PROGRAM WAS

Figure 7 – A PICTURE OF DEEP SPACE FROM THE HUBBLE TELESCOPE - GALAXIES GALORE

entitled *Cosmos*. It remains the most popular public television program in history and has been watched by 500 million people worldwide.

Sagan begins the book that accompanied the TV show by declaring: "The Cosmos is all that is, or ever was, or ever will be".

He goes on to offer a perspective that seems elegant, innocent, and to the scientifically-minded, obvious. But for those who believe in the Judeo-Christian understanding of origins, it offers immediate challenges to our way of thinking. Says Sagan:

> Today we have discovered a powerful and elegant way to understand the universe, a method called science; it has revealed to us a universe so ancient and so vast that **human affairs seem at first sight to be of little consequence.** We have grown distant from the Cosmos. It has seemed **remote and irrelevant to everyday concerns**. But science has found not only that the universe has a reeling and ecstatic grandeur, not only that **it is accessible to human understanding**, but also **that we are, in a very real and profound sense, a part of that Cosmos**, born from it, our fate deeply connected with it. The most basic human events and the most trivial trace back to the universe and its origins. This book is devoted to the exploration of that cosmic perspective.[1] [Emphasis mine]

Figure 8 – ASTRONOMER CARL SAGAN

On the surface, Sagan expresses a sentiment most can't help but echo. When we see pictures of the nebulae, the galaxies, even the planetary bodies in our own solar system, we are overwhelmed with wonder. But the question is whether *this sense of wonder* constitutes an adequate foundation to base our identity as a race, our reason for being here, and a knowledge of who God is and whether we are or are not aware of Him. It was enough for Carl Sagan and his present day protégé Neil deGrasse Tyson. But is it really enough to justify such optimism? Sagan's earnest passion was to find meaning in the universe and the place of human beings in it. His fundamental premise is

---

[1] Sagan, Carl. *Cosmos* (p. xvi). Random House Publishing Group. Kindle Edition.

INTRODUCTION

that the universe comprises a COSMOS – implying it has *a knowable design. Sagan's basis for humanity's purpose is our "connection" to the Cosmos.* For Sagan, humanity must explore and comprehend it, learning what it contains and how it functions. *We have meaning because we are a part of it.* But that is it.

HOW CAN WE JUSTIFY HUMANITY MATTERS AT ALL?

My point is this: Sagan's quest comes from a leap of faith – an optimism and enthusiasm that is neither logical nor realistic. While I may share Sagan's sense of the creation's grandeur and be awed by it, I don't draw conclusions about the meaning of humankind just because we find ourselves in a grand universe and we have a sense of its beauty. There has to be more to it than that. I appreciate the vastness of God's creation, His design, and am in awe of His power to create. Sagan and I agree that the creation is fabulous. But we have very different reasons for believing that humankind matters. I believe, and the authors writing with me believe, that we matter most because *God has placed us at the center of the story of the Cosmos.* And this premise of biblical faith provides the foundational point of this book. Thus, the symbol of Da Vinci's *Vitruvian Man* is used throughout this book to reiterate human beings (as God created them) are *at the center of the Cosmos.* This does not necessarily mean that the Earth is at the center of the Creation, although there is now some substantial scientific evidence that the Earth does lie at the center of the entire universe.[2]

My point is that God placed us here just as Psalm 8 testifies. The story arc of the Bible is the story of Adam and Eve placed in the Garden of Eden to be its caretakers. Humankind fell and that fall greatly impacted the Cosmos. The rest of the story arc then becomes how God redeemed human beings and will put them back into a paradisiacal setting in the New Jerusalem in a recreated Cosmos.

---

[2] This does not mean that the sun revolves around the Earth, but the location of our solar system in the vast expanse of the universe may be located centrally amidst the billions of galaxies that exist. See the film, *The Principle,* at http://www.theprinciplemovie.com/.

INTRODUCTION

In stark contrast, the atheist existentialist would demand much more than what Sagan offers. The atheist existentialist would **not** find meaning by *being awed by the wonders of the universe.* Sagan was awed. But the famous existentialist Jean Paul Sartre wasn't. An existentialist like Sartre who adheres to the principles of science without religious belief, declares that humanity has no hope or sense of purpose because there is no God to have made it so. Humankind finds itself here but cannot make any reference to God in order to certify human meaning, because *God does not exist.* There is no grand design. *The universe is NOT a Cosmos.* It is what it is. The universe remains meaningless because (in essence) human beings are *accidents*. We are merely the result of evolutionary principles and "chance". According to Darwinism, we are nothing more than a carbon-based canister of accumulated mutations – one mutation based upon another, and another, and another – until arriving at our present form. We have been *millions of years in the mill.* It's marvelous we wound up this way, but it means *nada.*

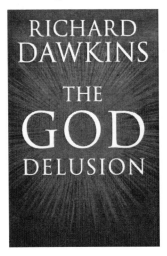

Figure 9 - THE GOD DELUSION (2006)

Then there are those like atheist Richard Dawkins who would make the evolutionary process itself "God".[3] Dawkins finds meaning in its "principles". But evolution with its principles of "natural selection" and "survival of the fittest" dismisses love and compassion. If God advances His creation through Darwinism, God is the devil. Humanity has no meaning.

---

[3] Citing blogger Albert Mohler (2011) "In *The Greatest Show on Earth: The Evidence for Evolution*, Dawkins sets out to present his most compelling case for evolution. He is — make no mistake — an ardent enthusiast for his argument. Seldom do we read a book written with such fervor and certitude, with an amazing amount of condescension and anger added to the mix, as well. "Evolution is a fact," he asserts. "Beyond reasonable doubt, beyond serious doubt, beyond sane, informed, intelligent doubt, beyond doubt evolution is a fact. The evidence for evolution is at least as strong as the evidence for the Holocaust, even allowing for eye witnesses to the Holocaust."

Note that this means, by obvious implication, that all objections to evolution are insane, unintelligent, and uninformed. Read his words carefully. Richard Dawkins is so bold as to assert that anyone who disagrees with him on such a controversial issue is insane, unintelligent, and uninformed". Of course, anyone in the world who holds to any religious belief would disagree.

INTRODUCTION

Altogether, atheism, existentialism, and Darwinism created despair. This despair dominated culture during much of the twentieth century. Just as Carl Sagan possessed a peculiar sense of hope in "awe", Dawkins had a peculiar sense of hope *in evolution.* But for the vast majority, evolution was "found out" to be much less than a hopeful dynamo advancing the species; it did not inexorably promote progress as Herbert Spencer (1820-1903) the English evolutionary philosopher had proclaimed. And as if that was not enough, optimism and hope collapsed once the view was established that *morality had no basis in reality.* The German philosopher Friedrich Nietzsche (who died in an insane asylum at the end of the nineteenth century), proclaimed *the death of God.* With His death morality died too, for God had always been the basis for morals. From then on, any meaning for humanity must be achieved when the *Übermensch* ("overman") would surpass conventional thinking (mostly Jewish ways of thinking according to Nietzsche) and establish *new values* – a transvaluation (or reevaluation) of values – based on stoic virtues lacking in compassion.[4]

Figure 10 – Philosopher FRIEDRICH NIETZSCHE

Therefore, without God, morality doesn't really exist. If we determine morality matters, it is our choice to assert it so. We can "talk morals" but we mustn't imagine that God exists to undergird them – that is, to ensure "the good guys win" or rewards await the moral person – for only

---

[4] "Elaborating the concept in *The Antichrist*, Nietzsche asserts that Christianity, not merely as a religion but also as the predominant moral system of the Western world, inverts nature, and is 'hostile to life'. As 'the religion of pity', it elevates the weak over the strong, exalting that which is "ill-constituted and weak" at the expense of that which is full of life and vitality." Retrieved from https://en.wikipedia.org/wiki/Transvaluation_of_values. This view of "compassionless love" would be exalted in the occult teachings of Aleister Crowley and *The Brotherhood of Saturn*, the most impactful German occult secret society that will figure into our story later. This "brotherhood" was formed in the same era as Nietzsche lived

INTRODUCTION

man, finite man (with his flaws) exists. But through his successor, Nietzsche argued the *Übermensch*, (mistranslated "superman"), can and would reinforce a new type of morality based upon a humankind that forsakes the delusion of God – a humankind that shoulders all responsibility.

Subsequently, unfettered from the old anachronism of religious faith, Germany would become the primary belligerent starting two world wars to *exert its will*. The *German "will"* must *overpower all others*. Henceforth, for Hitler and his accomplices, the Teutonic Race would guide humanity forward. Of course, we know that after twice failing to achieve its "Aryan destiny", Germany proved convincingly that when all compassion vanishes from human action, when morality is not grounded in the realm of absolute truth, *reality turns into nothingness*. Meaning is annihilated. Thus, the ruins of post-WWII Europe well illustrate the consequence of humanity serving as the sole basis for morals.

THE COSMOS BEGINS BY GOD *CREATING* IT & HUMANITY

So the point of all that was to say one cannot truly have a Cosmos nor propose any sort of substantive cosmology based on humanity alone.

The Bible says, *"In the beginning, God created the heaven and the earth."* (Genesis 1:1). As to how God created it exactly, whether with a Big Bang or through some other method less explosive, the Bible remains quiet. And yet, Moses the author of Genesis, moves rapidly through the steps of creation to underscore the making of humankind. "*So **God created man** in his own image, in the image of **God created he him**; male and female **created he them**.*" (Genesis 1:27) The Bible begins with much more than the mere fact that men and women find themselves in the universe – as part of it – owing their existence to it. Rather, the Bible declares that *the Cosmos begins with the Creator creating, and specifically, creating humanity*. It is this fact –*God created us* – *within which we find our purpose and can experience true meaning*. Three times in a single verse Moses underscored **God created us**. Moreover, as we soon learn, the fact that God created us was *in order to have a relationship with us*. This declares unequivocally *we matter because we matter to Him*! And since we matter to God, human beings are significant. But there is more to it still, much more.

INTRODUCTION

Although fallen creatures, **Jesus died for us.** In His vicarious death, God reinforces this yet again, He loves us. As Paul says, *"It is rare indeed for anyone to die for a righteous man, though for a good man someone might possibly dare to die. But God demonstrates His own love toward us, in that while we were yet sinners, Christ died for us."* (Romans 5:8, NIV) God created us for a high purpose – God redeemed us for this same purpose. We are to be *the custodians of the Cosmos.* These are the purposeful postulates upon which Christianity builds cosmology. We begin this book founded upon these premises.

WHY THIS BOOK WAS WRITTEN

The purpose of this book is to revise our reality, the awareness of Christians, concerning the Cosmos. We need this revision for two very important reasons: **One, we have been brainwashed with the secular, atheistic version of reality all of our lives**. We have been taught the Big Bang is the truth, as an incomprehensible and chaotic origin for all matter and energy. The brainwashing goes so far as to extol the virtues (I say this sarcastically) of an *inexorable expansion of the universe until it dies a death of entropy* despite the inherent contradiction that there exists an *evolutionary advance of our species "built into" the Cosmos.* However, we cannot stress enough that progressive evolution and death by entropy are *incompatible. Scientific atheism can't have it both ways.* Ironically, believers must be realists!

Likewise, we need to revise our reality because although we are a people of faith, **we have drifted far from the biblical premise that we live in a supernatural universe,** an understanding of the Cosmos where there are unseen superhuman creatures that directly influence our lives, as well as from the assumption that *God can and does intervene in the affairs of men and women.* God participates as an actor on the stage as well as serving as its writer and director, because He built the playhouse, He maintains it, and He produces the play itself.

We serve a God that is all-powerful and compassionate. He is not "the Burger King God" (i.e., "have it your way") for He has a greater plan for us. Our desires – our wishes and wants – may not be what is best for us. And yet, as Jesus said boldly, *"I am come that they might have life, and that they might have it more abundantly"* (John 10:10). Even in the Old Testament, God's sovereign oversight of our individual lives was understood. Below is but one example where the Psalmist sings praises for God's loving foreknowledge for each of us as individuals:

> [15] *My frame was not hidden from you*
> *when I was made in the secret place,*
> *when I was woven together in the depths of the earth.*
>
> [16] *Your eyes saw my unformed body;*
> *all the days ordained for me were written in your book*
> *before one of them came to be.* (Psalm 139-15-16, NIV)

NO MATTER HOW OLD YOU BELIEVE THE EARTH IS

However, the very next verse, Genesis 1:2 poses a controversy to all of us who believe in God as the Creator. It reads, *"And the earth was without form, and void; and darkness was upon the face of the deep. And the Spirit of God moved upon the face of the waters."* The issue where earnest believers may part company has to do with whether or not the earth was *initially created by God in a formless state, in a state of chaos and then further enhanced to today's form*, or whether it was a Cosmos *instantly* when fashioned by God's hand. The latter view asserts that something happened to change it *from a Cosmos to a chaos.* This view is better known as *the Gap Theory.* Those who believe in this point of view contend that something happened between Genesis 1:1 and Genesis 1:2 to cause the earth to become a "wasteland" (be formless – a void). Something transpired, presumably the fall of Satan and one-third of the angels from Heaven, that radically altered what had been created from the very first instant as a perfect Creation (or in steps not disclosed to us) to *render the Creation chaotic.*

On the other hand, the former view – *the Young Earth* position - supposes that the author of the Creation account offers *a pregnant pause* in his description. God made the raw elements of creation first and then He assembled them into their final form. There is no gap – but there were steps. No matter which view you hold, it asserts a huge cosmic premise. *God was the Creator – the world is not an accident. It was designed.*

INTRODUCTION

In the book before you, however, we will not take a particular position on this debate. We are *creationists*, generically speaking. We assert that God created the heaven and the earth. Thus, we embrace believers in three distinct camps regarding the age or nature of Creation: (1) those who advocate a "young Earth"; (2) those who support the notion of an

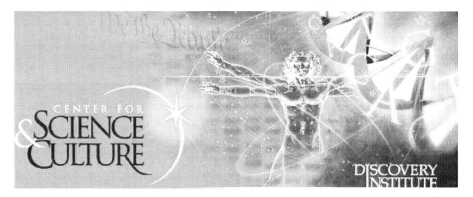

Figure 11 - DISCOVERY INSTITUTE, ADVOCATES FOR INTELLIGENT DESIGN

old earth that invokes "the gap" of time between Genesis 1:1 and 1:2 allowing for the possibility God's creation is billions of years old; and (3) those who simply believe in the *anthropic principle* [5]– that the universe was created in such a way that human life could exist... and as humans we could understand the complexity as well as unity within what is. For this third group, the age of the earth and its precise formation remains beside the point. *The creation was designed for us to live in it.* Scientific investigation underscores this fact again and again. Indeed, scientific investigation itself proves it was designed in such a way that we could observe and measure it. *The Discovery Institute,* a think tank in Seattle, Washington, focuses great minds on this endeavor. I am privileged to know a few of them.

---

[5] The *Oxford Dictionary of Physics* defines the anthropic principle in this way: "The principle that the observable universe has to be as it is, rather than any other way, otherwise we would not be able to observe it. There are many versions of the anthropic principle. The *weak anthropic principle* is specifically concerned with the conditions necessary for conscious life on earth and asserts that numerical relations found for fundamental constants, such as the *gravitational constant, have to hold at the present epoch because at any other epoch there would be no intelligent lifeform to measure the constants." A Dictionary of Physics (Oxford Quick Reference) (pp. 17-18). OUP Oxford. Kindle Edition.

In any event, all profess this truth: *God created us to be in fellowship with Him.* The *Westminster Catechism of the Reformed Church* proclaims,

> Q: **What is the chief end of man?**
> A: Man's chief end is to glorify God, and to enjoy him forever.

In all three of these cases (and probably in any permutation of these that the reader may hold), there resides a *shared belief in design* – that there was "a chief end in mind" when the Creator made us. In philosophy, it is called **teleology**, coming from the Greek *telos* ("end") and logos ("reason"), meaning an "explanation by reference to some purpose, end, goal, or function" (*Britannica Dictionary*). *God's end goal was to design and create by His Word a world for humanity to inhabit.*

What happens in this world ultimately is, as I say in my other books, "cosmically decisive". What happens here *matters*. And it matters to the entire Creation, not just our little blue planet on an outer spiral arm of the Milky Way Galaxy. At the risk of getting too theological, allow me over the next several pages to lay some important groundwork for our discussions that deals with the issue of "evil". *For it is also a key to the Cosmos.*

WHY IS THERE EVIL AND DEATH IN THE WORLD?

As the rest of the first chapter of Genesis unfolds, the emphasis is surely *that the creation is good.* It was not flawed from inception. It was not evil when God made it. Nor was there was anything latent within it that should cause the creatures God created to experience death, decay, or any manner of degeneration or diminishment. We were only instructed to be "fruitful and multiply and replenish the earth". No provision was made at the creation for the problem of sin and death, for the problem did not exist. God did not tell us, "You should fix any flaws you find in my construction – there is no warranty." *Caveat emptor* ("Buyer Beware!") NO! A hundred times, "NO!"

After working for six days (whether literal days or eons), the LORD God judged His creation *good* on *seven different occasions*, with the last "good" expressed as "very good". We believe that this was an emphatic statement to humanity *for all the ages (eons) to come*: "Know this… when I created the world, I created it without flaws."

INTRODUCTION

Additionally, the implication amounts to this: nothing evil exists innately or inherently in matter. The world is good. Your body is good. Reproducing is in fact your most essential mission! In contrast to many religions that suppose the seed of death was sown at the very beginning, God emphatically declares, "NO!" *It is good!* In contrast to those that might despise the body and assume that sexuality comprises a degrading act, God says, "I created physical attraction for a reason: Multiply!"

> *And God saw the light, that **it was good**.* (verse 4)
>
> *And God called the dry land Earth; and the gathering together of the waters called the Seas: and God saw that **it was good**.* (verse 10)
>
> *And the earth brought forth grass, and herb yielding seed after his kind, and the tree yielding fruit, whose seed was in itself, after his kind: and God saw that **it was good**.* (verse 12)
>
> *And God set them in the firmament of the heaven to give light upon the earth, and to rule over the day and over the night, and to divide the light from the darkness: and God saw that **it was good**.* (verse 17,18)
>
> *And God created great whales, and every living creature that moveth, which the waters brought forth abundantly, after their kind, and every winged fowl after his kind: and God saw that **it was good**.* (verse 21)
>
> *And God made the beast of the earth after his kind, and cattle after their kind, and every thing that creepeth upon the earth after his kind: and God saw that **it was good**.* (verse 25)
>
> *And God saw every thing that he had made, and, behold, **it was very good**.* (verse 31) [Emphasis added]

Christians believe that the reason God made it so clear "in the beginning" that *everything was good* was because something very bad was going to happen that would cause His creation and its creatures to "go sideways" – or in more proper theological wording – to *experience evil firsthand* by making an "unprogrammed first choice", a willful decision to go against the plan of God and disobey His rules. This decision was our decision. Adam and Eve made the choice to disobey God (and every person since then ratifies their choice). This outcome was certainly foreseen by God, but it was not foreordained or chosen by God. *Humanity chose to break the rules* – and so sin and death entered the Cosmos. The Bible says that all of creation has been subjected to this curse with an eye toward its redemption. (see Romans 8:20-22). *This state of things is a major cosmic principle too.*

While it will not be taken up here, there is a view in the philosophy of religion that evil must exist for good to be seen. That is to say, "We must have contrast or there would be no distinction." Judeo-Christian affirmations teach that this is a rationalization of those who knowingly or unknowingly are fooled if not consumed by evil – namely, humanity's most intellectual and esteemed thinkers. *Evil did not have to come into being to provoke the good.* And yet, God demonstrates that He can make good come from evil (Romans 8:28) *"to those who love Him and are called according to His purpose".*

We learn in the New Testament that God's creative act not only involved the Son of God, whom Christians identify as the Second person of the Trinity, but this Son was *the very person who brought the creation into being.* In the New Testament's Gospel of John, this Son would be called "the Logos" – *The Word of God.*

SOPHIA AND LOGOS – THE MIND OF GOD

However, let's recall that the notion of a *Logos* was originally a Greek concept predating Jesus Christ. It was a philosophy that God's Word, His reason and wisdom, existed throughout the creation and could – in part – *impart* wisdom. The Logos possessed power to illuminate or supply wisdom (in Latin, *Sophia*).[6] (This would be debated amongst the Church Fathers with some like Tertullian opposing philosophy while others like Clement (c. 150-220) and before him Justin Martyr (c. 105-165 A.D.) arguing that philosophy was useful as a means to provide truth even to those unaware of the Hebrew scriptures or the revelation of Jesus Christ.

But as it concerns the *Logos*: A Jewish religious philosopher, Philo of Alexandria (Egypt), a virtual contemporary with Jesus of Nazareth (Philo, b. circa 25 B.C. – d. 50 A.D., living 20 years before Jesus and 20 years after Him), had developed the notion of the *Logos* before John the

---

[6] This is the basis for "general revelation" and what is called "natural theology". But it has been the subject of great debate for 2,000 years, that is, whether this manner of revelation is able to bring any human being to a saving knowledge of God. See Romans 1:19-21.

INTRODUCTION

Apostle borrowed it for his gospel, circa 85 A.D. Many literary scholars of the Bible believe that John the Apostle first wrote his gospel to a targeted audience, a particular Jewish community in Northern Africa where many Jews had already relocated as part of the Jewish *diaspora* (i.e., dispersion) occurring before the time of Jesus and after the destruction of Jerusalem in 70 A.D. Philo's philosophy would be well known if not accepted by many of the Jews who lived there.

Philo sought to reconcile the Hebrew concept of God as Creator with the Greek notion of *permanent* matter (matter supposedly having always existed). Philo considered the *Logos* God's first creation – and still could be called His Son ("made not begotten" in contradistinction to the language of the Nicaean Creed). Therefore, Philo's *Logos* was not co-eternal nor co-equal with God. Indeed, he was known as a *demiurge*,[7] an

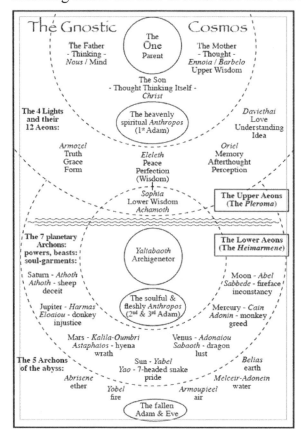

Figure 12 - THE GNOSTIC COSMOS
From Sethian Gnosticism c. 100-250 AD

---

[7] Wikipedia provides this concise description of the demiurge: "The word "demiurge" is an English word from *demiurgus*, a Latinized form of the Greek δημιουργός, *dēmiourgos* which was originally a common noun meaning "craftsman" or "artisan", but gradually it came to mean "producer" and eventually "creator". The philosophical usage and the proper noun derive from Plato's *Timaeus*, written c. 360 BC, in which the demiurge is presented as the creator of the universe. This is accordingly the definition of the demiurge in the Platonic (c. 310–90 BC) and Middle Platonic (c. 90 BC – AD 300) philosophical traditions." See https://en.wikipedia.org/wiki/Demiurge. The Brotherhood of Saturn, which we will address later, will speak of the demiurge and equate it with the ancient God, Saturn, substituting Lucifer for Jesus.

intermediary between God and creation. This view acknowledged that evil existed in the creation but this evil was not associated in any way with God nor could God approach evil. Therefore, since God could not directly touch or handle a tainted creation, an intermediary had to be created and placed "in between". God "emanated" a creative agent, indeed a series of "steps" between Himself and the creation (known as the *pleroma* or **all** the emanations of God assembled together). One such emanation which was less than perfect would bring the rest of creation into being. Like John's use of the term *Logos*, Paul would make use of the term *pleroma*, but he would correct its misconceptions and impart to it the *corrected* meaning – in accordance to God's revelation through Christ. Paul would later call this *pleroma*, the "fullness" of God – as in *all of God.* That was Christ.

This perspective remains vital to the idea of Gnosticism which existed before, during, and after Christianity began. It would be Christianity's greatest ideological opponent until atheism became the most dominant competitor to Theism in the twentieth century.

However, late in the twentieth century and into the twenty-first, *Gnosticism returned* to challenge orthodox Christianity. It appears today in the form of Theosophy, Luciferianism, and the New Age Movement. It has infiltrated almost all Christian theological systems, even supposedly conservative evangelicalism – especially in the form of the "Emergent Church"[8] movement in our day.

Because this is so important to define properly Christian cosmology, we will do a "deep dive" into Christology here in this Introduction. It serves as "prologue" for the rest of the book.

---

[8] What is the Emergent Church? It is an aggregate or combination of many things, where what you believe is secondary to your *predisposition to reject the standard beliefs, rituals, creeds, and structures of the "conventional" church.* "The emerging / emergent church movement falls into line with basic post-modernist thinking—it is about experience over reason, subjectivity over objectivity, spirituality over religion, images over words, outward over inward, feelings over truth. These are reactions to modernism and are thought to be necessary in order to actively engage contemporary culture." From http://www.gotquestions.org/emerging-church-emergent.html.

INTRODUCTION

CHRISTIANITY'S LOGOS IS NOT PHILO'S LOGOS

Taking into account the notion of *Logos* that his audience understood, John employed the term *Logos* too, but gave it new and profound meaning. We know this because we know of Philo, what he taught, and we can read between the lines to detect the clarification inherent within John's wording in his gospel's opening chapter:

> *¹ In the beginning was the Word (LOGOS), and the Word was with God, and the Word was God.*
>
> *² The same was in the beginning with God.*
>
> *³ All things were made by him; and without him was not any thing made that was made.* [Emphasis added]

InterVarsity Press' *Pocket Dictionary for Apologetics and Religion* provides this definition of **Logos**:

> Greek term for "Word" or "Reason," used in the prologue to John's Gospel: "In the beginning was the Word." The Logos is thus a term for the eternal Son of God, the second person of the Trinity, understood as the agent of divine creation and the one who "illumines," or enlightens, human beings. Many early church fathers used this concept to justify a positive attitude toward Greek philosophy on the grounds that those who lacked the biblical revelation could still attain some truth because of the operation of Christ as the Logos in them.[9]

The idea that the Logos could create and yet exist within the Godhead *was distinctive to New Testament Christianity.* Given the context of its beginnings, Christianity had to embrace the goodness of God's creation, and yet acknowledge that evil existed in the world. Therefore, it must draw a line between the physical creation and the reality of sin and death. Evil was not "built into" matter; but was a later interloper, like a virus that infected the creation. Evil was introduced not by God Himself, but by one or more of His creatures.[10] The Hebrews believed that God was one person without

---

[9] Evans, C. Stephen. *Pocket Dictionary of Apologetics & Philosophy of Religion: (The IVP Pocket Reference Series)* (pp. 70-71). Downers Grove: InterVarsity Press. Kindle Edition.

[10] When the Old Testament speaks of God bringing evil or causing evil, it is my perspective that this "evil" should be understood as an effect, not as a cause. Or more specifically, if something bad happens to a person, "evil has happened" and in the context of God's providence, God

INTRODUCTION

any distinguishing of persons within the Godhead (no concept of a "Trinity"), although God appeared from time-to-time to human beings in the Old Testament. In doing this, His appearance was explained as the *Angel of God*. And yet, the *Angel of God and God* in Hebrew passages in the Old Testament might both appear at the same time and same place. (We will take up this "duality" and ultimately the "tri-unity" of God in more detail in due course). [11]

First, the notion that a divine entity other than the "highest" God could create the world was certainly entertained before the time of Christ. Greek religious philosophy had developed mystery religions and philosophies that we would generally equate with Gnosticism.

Next, Gnosticism posited a *demiurge* as the creator. But Christianity would reject this idea of a "middleman". In developing its scripture, the apostles of Jesus would complete an overhaul of the cosmology of the ancient world. John and Paul were the primary authors who **revised reality.** As noted before, this must be done again today because so many premises of the Bible are now violated – *how Christians think about the physical and spiritual worlds has shifted.* We must return to biblical ideas to experience *God in all His loving power* and *to defeat the enemy of our souls*.

Paul would take pains to express the nature of the Logos, the Christ, in Colossians, chapter 1, where he expresses a "cosmic Christ":

> [15] *Who is the **image** of the invisible God, the **firstborn** of every creature:*

---

has allowed evil to come to pass. But this should be understood metaphorically not literally. Admittedly, the wording of some scriptures seems to state otherwise.

[11] Examples of when the ***angel of the Lord*** appeared in the Old Testament:

Jacob wrestling with the Angel of the Lord while seeking a blessing (Genesis); The Angel of the Lord testing Abraham's faith as Abraham prepares to sacrifice his son Isaac (Genesis); The Angel of the Lord comforting Hagar as she wanders through the desert with her son Ishmael (Genesis); The Angel appearing to Moses in the burning bush (Exodus); The Angel of the Lord leading the Hebrew people toward the Promised Land (Exodus); The Angel of the Lord encouraging Gideon before a battle (Judges); The Angel of the Lord confronting Balaam about his mistreatment of an animal (Numbers). See http://angels.about.com/od/Famous-Archangels/p/Meet-The-Angel-Of-The-Lord.htm.

*¹⁶ For by him were **all things created**, that are in heaven, and that are in earth, visible and invisible, whether they be thrones, or dominions, or principalities, or powers: **all things were created by him, and for him**:*

*¹⁷ And he is **before all things**, and by him **all things consist**.*

*¹⁸ And he is the head of the body, the church: who is the beginning, the **firstborn** from the dead; that in all things he might have the **preeminence**.*

*¹⁹ For it pleased the Father that **in him should all fullness dwell**.*

Several points should be underscored. Permit me to "exegete" these five verses to dig out the meaning as it is so vital to our study:

1. Christ is the exact **image** of God (v. 15, *eikōn in the Greek* from which our word *icon* comes) because it pleased the Father that in Him (Christ) should all the fullness (*plērōma*) of God dwell. The implication: Christ was not partially divine – He was fully divine. There are no "gradations" or emanations between God the Father and His Son. The Son was not in any sense missing an aspect or element of God. In using *eikōn*, Paul may have been distinguishing the essence of Christ from an *aeōn* (a word that sounds similar in the Greek, but was employed by the Gnostics and means something quite distinct).[12] The idea of *aeon* connects to an era (eon) in English. *Aeon* links the creation in both space and time, but we will not discuss that notion here. We need only make note that *aeons* are "steps" or different levels from the "most divine to the least divine" in the Gnostic system.

2. Christ is the **firstborn** (v. 15, *prōtotokos*) of every creature (or creation) in the sense of preeminence, coming before every other creation. As "he is before all things" (*pro* in the Greek meaning "prior to"). Explicitly stated in v. 18: "that in all things he might have the **preeminence**" (*prōteuō*), which is the only time this verbal form of the adjective *prōtos* is used in the New Testament. *Prōtos* conveys "standing in first place all alone". *Prōteuō*, as a verb would emphasize an active status (*ongoing*) as opposed to a passive state (*already*). *Christ stands alone.* The King James mistranslates the

---

[12] "This source of all being is an Aeon, in which an inner being dwells, known as Ennoea ('thought, intent', Greek ἔννοια), Charis ('grace', Greek χάρις), or Sige ('silence', Greek σιγή). The split perfect being conceives the second Aeon, Nous ('mind', Greek Νους), within itself. Along with male Nous comes female Aeon Aletheia ('truth', Greek Αληθεια). These are the primary roots of Aeons. Complex hierarchies of Aeons are thus produced, sometimes to the number of thirty. These Aeons belong to a purely ideal, noumenal, intelligible, or supersensible world; they are immaterial, they are hypostatic ideas. Together with the source from which they emanate, they form Pleroma ('region of light', Greek πλήρωμα). The lowest regions of Pleroma are closest to darkness—that is, the physical world." See https://en.wikipedia.org/wiki/Aeon_(Gnosticism)

word as if it were a *noun* instead of a *verb*, "holding the first place *status*, i.e., possessing the "blue ribbon". Hence, it misses an important nuance to Paul's meaning.

3. God created all **things** (in Greek, *pas* meaning "everything or all things from top to bottom") **through Christ** "for by him were all things created" whether in heaven or in earth (i.e., throughout the entire creation), both visible things and invisible things. Then Paul lists entities that may reflect the "divine counsel" referenced in the Old Testament – entities that exist "above the angels". These appear to be more than "ranks" but rather "positions" ordered from highest-to-lowest, yet all leaders in the divine scheme of things or in the "organization chart" of creation. Paul lists four levels of "authorities": **thrones** (*thronos* – a seat signifying a head of state as a kingship); **dominions** (*kyriotēs* – from *kyrios* a lordship, a step below a king, as a lord in beneath a king in an English rank); **principalities** (*archē* – someone in a leadership position, a magistrate, from *archomai,* the first one to do something, to be first in line); and finally, **powers** (*exousia* – the one who has the authority to make decisions, who has liberty to do as he or she pleases). Paul takes the time to list them all to point out the various rankings implied in the *aeons* are all subservient to Christ for *He created every one of these.*

4. In New Testament Christianity, we must place Christ *at the center.* As Paul said, "he is before all things, and by him all things **consist**" (v. 17 *synistēmi* in the Greek which means "to bring together, compose, connect, unite the parts into a whole.") Christ sustains the creation actively moment-by-moment. Without Him, the creation would not continue to exist. In Theology, Christ constitutes the underlying "ground of being".

EVIL IS MORE THAN THE DARK SIDE OF THE FORCE

Besides a "prologue" in respect to Christology as a foundation to our revision of reality – our cosmology, we must go just a few steps further regarding *the nature of evil* and provide additional context. If we don't, some philosophers, theologians, and agnostics will dismiss this study as *ignorant of the classic issues in Theology.* We wish to acknowledge we are quite aware of the issues and our discourse transpires in light of them. To address the issue, we provide this summary on "evil" for all readers.

The problem of evil is generally stated as a dilemma: "How can God be all good and all powerful and there be evil in the world? He is either evil for allowing it to exist, or worse, He created evil. Alternatively, He cannot be

INTRODUCTION

omnipotent for horrid things happen in our world. Some atheists argue that if such a God existed, there would be no evil, since God would both want to eliminate evil and would be able to do so. An argument that evil is logically incompatible with God's reality forms the logical or deductive form of the problem. An argument that evil makes God's existence unlikely or less likely is called the evidential or probabilistic form of the problem." [13]

In Theology, a simple question captures the essential issue, "Why does evil exist?" The answer of course is not so simple. The response to the question attempts to explain why God remains "all good" and "all powerful". It is known as a ***theodicy***. IVP's *Pocket Dictionary of Apologetics and Philosophy of Religion*, provides this succinct definition:

> An answer to the problem of evil that attempts to "justify the ways of God to man" by explaining God's reasons for allowing evil. Two of the more important theodicies are the "soul-making theodicy," which argues that God allows evil so as to make it possible for humans to develop certain desirable virtues, and the "free will theodicy," which argues that God had to allow for the possibility of evil if he wished to give humans (and angelic beings) free will. Theodicies are often distinguished from defenses, which argue that it is reasonable to believe that God has reasons for allowing evil even if we do not know what those reasons are.[14]

The challenge posed for Cosmology also involves *metaphysics* (the "really real" or ultimately real elements of the Cosmos). Specifically, the conundrum comprises "the *dualism* dilemma": Is all reality subsumed in being one or the other: is all that is either good or evil?

All religions attempt to explain the universe using the terms *good* and *evil*. Plus, critics consider "dualism" bad form and an intractable problem to avoid at all costs.

So do Jewish and Christian theological worldviews find themselves trapped into a dualist perspective in which *all that is* comprises either good or evil? Asserting that Satan and God constitute exact opposites *would explicitly*

---

[13] Evans, C. Stephen. *Pocket Dictionary of Apologetics & Philosophy of Religion: 300 Terms & Thinkers Clearly & Concisely Defined* (The IVP Pocket Reference Series) (p. 42). InterVarsity Press. Kindle Edition.

[14] Ibid., (p. 114).

convey dualism. And we don't hold this position. The IVP reference guide says this about dualism,

> For example, ancient *Manichaeism* was a form of dualism postulating two equal but opposing divine realities, a good power of light and an evil power of darkness. Theism has a dualistic dimension in that it makes a clear distinction between God and the created order, between the infinite and the finite.[15]

Make note that Gnosticism speaks *in favor of dualism* and in so doing, sides with religious philosophy over biblical doctrine. C. S. Lewis provides a most intelligent response to the challenge outlined here. I will cite only a morsel of his "solution" and refer the reader to the full discussion within his book, *Mere Christianity.* Says Lewis:

> What is the problem? A universe that contains much that is obviously bad and apparently meaningless, but containing creatures like ourselves who know that it is bad and meaningless. There are only two views that face all the facts. One is the Christian view that this is a good world that has gone wrong, but still retains the memory of what it ought to have been. The other is the view called *Dualism*. Dualism means the belief that there are two equal and independent powers at the back of everything, one of them good and the other bad, and that this universe is the battlefield in which they fight out an endless war. I personally think that next to Christianity Dualism is the manliest and most sensible creed on the market. But it has a catch in it.[16]

Lewis goes on to point out what "the catch" is. We can't simply attach "good" to God and evil to "Satan". Nor can we state that *we have a preference for one or the other* because that isn't really certifying something to be truly "good" or "bad". Furthermore, if we say that there are *rules to make one right and the other wrong*, then the *rules* for what is right or wrong wind up being established as the "higher God" than either of the two powers presumed to be the equivalent of "good" and "evil". Consequently, we don't resolve the dilemma this way. We must not talk about good and evil "existing" as if they are realities unto themselves, locked in eternal opposition

---

[15] Ibid., (pp. 36-37).

[16] Retrieved from the site of Randy Alcorn, see http://www.epm.org/blog/2012/Sep/24/c-s-lewis-good-and-bad

INTRODUCTION

like "Yin and Yang". Instead, we must talk about *persons or entities possessing both power and will* to make decisions and put their decisions into motion within the creation.

Having said that, I am now able to emphasize that *much of this volume will deal with entities, both human and non-human* that in fact exist in the visible or invisible worlds of the Cosmos. Much of the story we will tell resides in describing the relationships between these entities. These relationships are *fundamental to the Cosmos*, they are *dramatic*, and they are *some of the best kept secrets of Judeo-Christian thought.*

Therefore, we will not get lost in a *dualist dilemma.* But we will dig deep into a discourse focused on two types of beings/entities. First, we will consider persons who love, nurture, and care about the creation and who choose to side with the Creator, the only true God who we know as Yahweh as well as His co-eternal Son who actually did the creating, Jesus Christ. Secondly, we will discuss beings/entities that are persons who despise, destroy, and seek to *transmogrify the creation including us - humanity* (like a type of negative alchemy), into something other than what the Creator originally designed and fashioned it. These sentient creatures are persons both human and non-human.

Now, as it concerns God and Satan: First, we affirm that the God Yahweh is all good AND all powerful. Secondly, Satan is a created being, a person who opposes God. He is not equal in power nor, as Mormonism supposes, is he a son of God like Jesus is (as if they were brothers) caught in a sibling rivalry. *Jesus is co-eternal and co-equal with God.*[17]

Having completed our "prologue", we will return to the *Introduction*.

---

[17] Mormon Cosmology is a fascinating but confusing amalgamation of Freemasonry, Gnosticism, Christianity, Hinduism, and the "gospel of Extraterrestrials". It is highlighted by the assertion that God's triune nature involves each of the members of the Trinity possessing bodies, including having flesh and bones. It remains quite foreign to biblical cosmology and anathema.

INTRODUCTION

NEW PARTICLES AND DIMENSIONS TO THE CREATION

Much of this book will address the fabric of the universe, how it is composed, and why it behaves the way it does. There has been a great deal of talk lately about space/time, about whether there are really ten dimensions plus time (as an eleventh dimension), or whether time is really a dimension at all (as someone said, "It's the only dimension to which you cannot point or tell someone else what direction it's going").

How many dimensions are there? We actually won't propose a specific answer to that question. We offer several ideas. But don't worry. We won't develop mathematic formulas or theorems attempting to prove the exact number of dimensions in the Cosmos;[18] however, we will discuss why the universe involves much more structure than just "height, depth, and width". And no topic seems more exciting to most readers who've likely picked up this book than whether science is about to "open a portal to another dimension" and wonder if something or someone will come through the portal when its propped open.

Much of current science seems obsessed with discovering new particles and explaining why 96% of the universe cannot be seen, measured, or described in terms of "what it is", rather commenting only on what it is not. I am speaking here of "dark matter" and "dark energy". We will challenge whether or not these notions (which have become part of the standard view of the Cosmos) really exist at all. We will also elaborate on an alternative theory gaining traction today called *The Electric Universe*. While it has not been demonstrated equally to the satisfaction of all the authors contributing to this book (that it constitutes a better and complete model for why the universe works the way it does), this alternative model does offer some answers that the "standard model" doesn't. We will talk about these issues in layman's terms in this book.

---

[18] It should be noted that the notion of 10 dimensions was conjured up with the concept of "String Theory" – that all that is at the lowest level are strings of different lengths, different vibrations, and different shapes. However, String Theory has fallen on hard times for all except those who invested 20 years of their careers seeking to prove it. But the view that there are 10 dimensions appears now to be no more than residue from the once vaunted view. There is considerable debate on this still. Here is one article that summarizes it: https://www.theguardian.com/science/2006/oct/08/research.highereducation.

INTRODUCTION

Oh, by the way, the odd line you see across the bottom of the page is a picture of a Birkeland Current, derived from a photo taken by the Hubble Telescope. These currents are spirals of electricity that play no small part in the story of the Electric Universe.[19] I will offer the rationale here that the Electric Universe contends the electricity and magnetism (electromagnetic powers) have far more to do with the Cosmos than gravity. This issue – Einstein vs. Tesla – will also be taken up later.

Lastly, on the topic of particles and dimensions, one of the highlights of the book will be Anthony Patch's presentation on the Large Hadron Collider (LHC) at CERN. I promise that his discussion on what is happening there will blow your mind and likely alter your eschatology too.

In contrast, Josh Peck offers a good deal of practical insight into the "multi-verse" – realizations about the Cosmos that expand our awareness of the creation and how the Creator sculpted it as a habitat for humanity, the other creatures living within it, and the non-human beings that exist beyond the three dimensions of space *and time* (as we experience it) in what is generally the invisible "side" to the Cosmos. Josh's concepts and his "way with words" will help the reader come to terms with these additional dimensions and why they ultimately promise to impact our lives in practical ways. His approach and demeanor are vitally important to provide a much needed balance to my rigorous style which borders on the "taking oneself too seriously" and being a bit too careful in choosing words. Sorry if that frustrates you a bit.

TECHNOLOGY AND THE DESTINY OF HUMANITY

Speaking of eschatology, this book has been written by the authors from the perspective that we are living in *the very last of the last days*. Writing on the topic of Cosmology would hold little interest to us (and we venture to you the reader), if the context of our discussion wasn't relevant to what the Bible predicts about the last days and what the intersection between

---

[19] "A Birkeland current usually refers to the electric currents in a planet's ionosphere that follows magnetic field lines (i.e. field-aligned currents), and sometimes used to described any field-aligned electric current in a space plasma. They are caused by the movement of a plasma perpendicular to a magnetic field." See www.plasma-universe.com/Birkeland_current.

science and theology means to the matter of what is soon to happen in the "real world" in which we live.

Consequently, Gonzo Shimura addresses the implications of the technologies of physics and transhumanism in two of his essays (to one of which I also contribute). His is especially eschatological. His compositions for this book address several major concerns:

- The possibility that humanity will be altered genetically and through the insertion of mechanical (digitized) devices into our bodies. This is the stuff of horror movies. Gonzo has tracked the progress of these ideas over a number of years and is well-versed to speak to these issues

- The likelihood humanity will be dramatically altered – even achieve immortality – with unintended consequences. This has been a subject taken up by many authors both religious and secular. Ray Kurzweil's *singularity* predicts humankind will confront a crossroads in 2045 that will challenge the usual conception of our destiny inexorably marching toward greater human potential. Instead, we may be dwarfed by machines that outperform us intellectually and physically. What it means to be human will undergo drastic revision. If this occurs, the Cosmos may no longer be the safe haven we presume it is at this moment in time. Despite the ominous forecast for the diminution of humanity's well-being (unless it is supplemented by the transhumanist's remedies to our limitations), Gonzo and the other authors remain opposed to transhumanist "fixes" that are more likely to generate unintended and unwanted consequences. In his discussions, Gonzo will make this quite evident.

- Likewise, physicist and cosmologist Stephen Hawking warns that humanity must come to grips with the irrationality of a universe possessing black holes that could annihilate reality. He has explored anew the matter of determinism and how humanity will deal with a dismal destiny where what happens to our species escapes our control. Here as well, Gonzo confronts this dark possibility with a characteristic hope derived from the themes of the Bible, perspectives that promise our Cosmos stands on the cusp of a radical transformation toward restoration and recreation, rather than destruction and death.

Lastly, all the authors address the possibility that humanity faces another "Tower of Babel" moment, in which technology is being wielded behind the scenes by power-player elites to the detriment of the vast majority of

INTRODUCTION

the world's population. New weapons of mass destruction, new schemes to rid the world of over 90% of its population (as a means to protect the world for the rich "ennobled" class), and secret connections established with purportedly beneficent extraterrestrials who will evolve humanity to a better future, are all increasingly expansive threats breaking in on the awareness of many.

The extent of these problems are sweeping to say the least they could be literally described as *cosmic in scale* – hence, their inclusion in the scope of this book. Indeed, how these issues have become interwoven with Bible prophecy constitutes a large portion of what we address in this first volume of what is intended to be a two-volume set.

Finally, toward the end of this book, I provide a chapter addressing the possibility that the battle between Yahweh and His opponents stretches beyond the earth, our solar system, and certainly beyond the usual dimensions of space and time. It includes weapons of war transcending our wildest imaginations. It begs certain questions we will talk about in the next volume. And as a bridge to the next book, we begin in this volume to discuss the possibility that to understand the Cosmos *we must throw away what we know about our Sun, our Moon, and how the solar system* works; that it, how it may have been organized at its beginning until recent catastrophes radically altered it. In so doing, we will then be better able to discover the actual connections between the planets, mythology, anthropology, and the secret societies that have and continue to influence the arc of human history, a pathway leading us toward oblivion if the LORD God does not call a halt to humanity acting in blatant disregard to His will.

This is our agenda. We best get to it.

INTRODUCTION

HOW THE NEPHILIM CHANGED THE COURSE OF OUR COSMOS

Josh Peck with S. Douglas Woodward

FAR OUT DUDE

OVER THE PAST DECADE OR TWO, THOSE WHO HAVE FOLLOWED BIBLE PROPHECY HAVE BEEN INUNDATED WITH A WHOLE NEW STRAIN OF FAR-OUT INFORMATION, SO STRANGE IN FACT THAT THOSE WHO might have dared tout it two decades ago would have been assigned to the loony bin. Not today.

Topics like UFOs, extraterrestrial visitation to our planet, alien abductions, alien breeders, "the watchers", and fallen angels breeding with human beings creating a hybrid race called "the Nephilim" all have become common conversations in circles where eschatology is taken seriously. In fact, interest in these oddities has become so intense, it has almost overwhelmed the traditional study of futurist eschatology. Most Bible prophecy scholars now include these esoteric subjects as a key part of the story of biblical history and as an explanation for the paranormal phenomenon so much a part of our "pop" culture. The highly regarded evangelical scholar Michael S. Heiser even argues that the existence of the Nephilim is indispensable to understand the scripture and the plan of God for Israel.[1]

Far from being "fringe" teachings, these notions comprise the core of a proper understanding of the Cosmos. In effect, we can call what has happened a "sea change." For Christians, the Cosmos has been transformed.

---

[1] See Michael S. Heiser, *The Unseen Realm.* Bellingham, WA: Lexham Press, 2015.

The explanation eschatology experts offer regarding why these formerly fringe topics are all the rage today builds upon the notion of **the great deception in the last days.** 2 Thessalonians 2:7-12 teaches:

> *For the mystery of iniquity doth already work: only he who now letteth will let, until he be taken out of the way. And then shall that Wicked be revealed, whom the Lord shall consume with the spirit of his mouth, and shall destroy with the brightness of his coming: Even him, whose coming is after the working of Satan with all power and signs and lying wonders, And with all deceivableness of unrighteousness in them that perish; because they received not the love of the truth, that they might be saved. And for this cause God shall send them strong delusion, that they should believe a lie: That they all might be damned who believed not the truth, but had pleasure in unrighteousness.*

Figure 13 - AN ALIEN ENGINEERS HUMANITY IN THE MOVIE PROMETHEUS

Such lying wonders today demand "biblical contexting" (to coin a phrase) if Christians are to understand "what is really going on". We know that there were ancient accounts of breeding between human and "gods" among many primitive cultures located at divergent places around the globe. We naturally ask, "Why would these megalithic people have knowledge of star systems in outer space?" And even more curious among indigenous Americans are reports (throughout their history, both oral and inscribed in pictographs) of strange visitors claiming to have come to Earth from another planet. How could these accounts be reconciled to what the Bible teaches about ancient history and a "Bible-centric" view of the Cosmos?

For those of us who believe in supernatural evil, namely the power we call Satan, it doesn't take much imagination for us to believe "that old dragon" (Revelation 20:2) is the source of such disinformation – not only to deceive ancient humankind, but also to deceive us today… since so much science fiction has woven these fantastic ideas into the fabric of myriad mysteries we witness about outer space now. We think of H.P. Lovecraft, Ray Bradbury, Rod Serling, or more recently movies like *Prometheus* from the *Alien* franchise, bringing Lovecraft's notion of ancient aliens visiting Earth aeons ago (i.e., "The Cult of Cthulhu") to the big screen – scenes impossible to distinguish from reality.

In fact, since he has the capability to do so, Satan would be hard-pressed to construct a better plan to deceive the whole world into worshipping him and his minions, than this weird take on the Cosmos. Hollywood has indeed created a modern religion with grand appeal because it combines science, science fiction, and an explanation for our genesis. *Transformers, Independence Day, Aliens*, multiple episodes of *Star Wars*, and *Star Trek* – the list goes on and on. The indoctrination conducted on the silver screen (and the "boob tube") has had a vast impact on how we think about the Cosmos today.

With this vision in the background, just imagine how the world would react if alien spacecraft actually appeared and made contact with world leaders live on TV. With the whole world watching this mesmerizing "reality", who would be willing to contradict **anything** the alien beings proclaim?

Crazy as it may seem, if this set of circumstances were to actually happen it wouldn't be long before Satan and his minions would have the "earthlings" fully hoodwinked and within their grasp.

SONS OF GOD

But the story we four authors are here to tell (Josh and Doug in this chapter) all *begins with those beings identified as the sons of God*. This may seem like a strange beginning, but it is this strangeness that launches us onward toward a revised understanding of biblical reality, or, in other words *our reality the way the Bible sees it*. We have witnessed the identifier *sons of God* generate considerable controversy inside the Church throughout the ages and even more in our day. We have heard just about every definition imaginable to explain this phrase. The *sons of God* have been identified as "sinful men", "descendants of Cain", "demons", "sons of Seth", and even "aliens from another planet". That is why we maintain it's best to look at the scripture across multiple passages to see what the whole counsel of God has to say about who they were (and we believe still are).[2] To rightly understand the biblical viewpoint, we should begin at the beginning. For that, we go to Genesis 6:1-2, which reads,

> *And it came to pass, when men began to multiply on the face of the earth, and daughters were born unto them, That the sons of God saw the daughters of men that they were fair; and they took them wives of all which they chose.*

---

[2] Modern-day commentaries on these passages strenuously deny that the "sons of God" could be angelic beings, and refer instead to the standard liberal view that the writer of Genesis was decrying the intermarriage of the godly bloodline of Seth with the ungodly bloodline of Cain. *The Keil Delitzsch Old Testament Commentary* would be a case in point:

> The question whether the "sons of Elohim" were celestial or terrestrial sons of God (angels or pious men of the family of Seth) can only be determined from the context, and from the substance of the passage itself, that is to say, from what is related respecting the conduct of the sons of God and its results. That the connection does not favour the idea of their being angels, is acknowledged even by those who adopt this view. "It cannot be denied," says Delitzsch, "that the connection of Genesis 6:1-8 with Genesis 4 necessitates the assumption, that such intermarriages (of the Sethite and Cainite families) did take place about the time of the flood (cf. Matthew 24:38; Luke 17:27); and the prohibition of mixed marriages under the law (Exodus 34:16; cf. Genesis 27:46; Genesis 28:1) also favours the same idea." See http://biblehub.com/commentaries/kad/genesis/6.htm.

This verse does not directly tell us who the *"sons of God"* are. To define these *"sons of God"*, we best employ the Reformation principle of allowing scripture to interpret scripture (in Latin, *scriptura scripturae interpres*). The book of Job uses the expression *sons of God* twice. First we read in Job 1:6: *"Now there was a day when the sons of God came to present themselves before the LORD, and Satan came also among them."*

Secondly, we read in Job 38:4-7:

> *Where wast thou when I laid the foundations of the earth? declare, if thou hast understanding. Who hath laid the measurements thereof, if thou knowest? Or who hath stretched the line upon it? Whereupon are the foundations thereof fastened? or who laid the corner stone thereof; When the morning stars sang together, and all the sons of God shouted for joy?*

This is a prime example of how we must allow the Bible to define its vocabulary for us. Comparing one scripture to another supplies a clear understanding that these *"sons of God"* are the angels in heaven and not "the sons of Seth" down on earth. Those that disagree with this assessment typically do so from an anti-supernaturalist bias (that is, they don't believe in angels and demons). No wonder they disagree with most everything else evangelicals believe.

Two points support the conclusion that we are talking about angels here.

First, these "sons" presented themselves before the LORD in heaven. Know any living human being that has done that lately?

Secondly, they were alive and accounted for at the creation of the world. However old the world is, the angels predate the earth. Unless we believe in polytheism and the **eternal** existence of many distinct and important gods, this could only mean these sons of Gods were created beings.[3] As to their standing with God, this concept conveys that they are direct creations of

---

[3] Michael S. Heiser takes up the notion of "many gods in council with Jehovah" in his brilliant book, *The Unseen Realm*. There are numerous passages that discuss this. One of those is Psalm 82:1, *"God standeth in the congregation of the mighty; he judgeth among the gods."* And verse 6, *"I have said, 'Ye are gods; and all of you are children of the most High.'"* These elohim are known as the "Divine Counsel", are discussed briefly here, and more in depth in a subsequent chapter.

God (as was Adam and all his progeny – that is, you and me).[4] On the other hand, it doesn't specify whether they were in good standing with God. We can safely presume some were and some were not.

We should make a special note that there is what the Bible calls, a "host of heaven", a council that the God of the Bible holds for executing His plan. We read about it in Psalm 82:1,6 and also in Psalm 89:5-8:

> *The heavens will praise Your wonders, O LORD;*
> *Your faithfulness also in the assembly of the holy ones.*
>
> *For who in the skies is comparable to the LORD?*
> *Who among the sons of the mighty is like the LORD,*
>
> *A God greatly feared in the council of the holy ones,*
> *And awesome above all those who are around Him?*
>
> *O LORD God of hosts, who is like You, O mighty LORD?*
> *Your faithfulness also surrounds You.*

While it's beyond our scope here in this chapter to expound on this "host of heaven", suffice it to say that before humans were created, entities called *elohim* in Hebrew existed and Scripture references them in many places. Perhaps they should be categorized as either cherubim and seraphim – but we suspect that they are distinct from the "average, run-of-the-mill angel". Theologians debate the exact timing of their creation, but clearly they existed before homo sapiens. No matter who these entities are, they appear to be highly intelligent and powerful. And in the New Testament the "host" likely corresponds to the "powers and principalities" (Ephesians 6:12) depicted by Paul.

It is intriguing that they are called *elohim* in the Hebrew. But before you think us polytheists, we underscore we are "modified monotheists" – that is, we are orthodox Trinitarians. Further explanation will be deferred until Gonzo and Doug take this up in a later chapter.

---

[4] Understanding this point brings added meaning to Jesus taking on the name, Son of Man. That title emphasizes that his DNA is human. He was not a hybrid. Satan's plan to corrupt human DNA appears NOT to have ended with the flood, but persists even today.

GIANTS

When we look at the Genesis 6 account, we discover these *sons of God* saw the *daughters of men* and took them for wives. Their offspring were giants. But unless we accept the so-called Sethite Theory (that two different human bloodlines combined creating offspring of great size, literally exemplifying "hybrid vigor" to the "n^{th} power"), something very peculiar took place when they mated. Shortly after the match made in hell transpired, Genesis 6:4 states, *"There were giants in the earth in those days; and also after that, when the sons of God came in unto the daughters of men, and they bare children to them, the same became mighty men which were of old, men of renown."* Just in case it isn't obvious, in this passage the biblical writer (conservatives believe it was Moses) just told us that those Greek mythological figures, the Titans and Olympians had **a biblical basis** (many whose names learned in High School English no doubt we've mostly forgotten by this point in our lives).

Figure 14 – THE FALL OF THE REBEL ANGELS BY HIERONYMUS BOSCH
(CIRCA 1450-1516)

The demigods were not just plucked out of thin air –real history stands behind these myths. The evangelical intellectual Francis Schaeffer asserted this very same point years ago.

The word *giant* is translated from the Hebrew word *nĕphiyl,* meaning "bully" or "tyrant", which appears related to the word *napal,* meaning, "to fall". Isaiah 14:12 uses the same word *napal* for our English word *fallen* to describe Satan. We recall the common notion that mankind had a "fall" in the garden of Eden. This fall was *not* tripping over the root of a tree growing there – it was much more cataclysmic than that. It was a descent from one level of existence to another, from a particular exalted moral status to a much lower place in the grand scheme of sentient beings.

As the reader likely knows, the standard word used today for these fallen tyrants, or giants, is *nephilim.* There have been many arguments advanced and a few word studies offered up concerning this term and, though we will refer to the giants as *Nephilim*, just keep in mind that according to the Strong's Exhaustive Concordance, the actual Hebrew word is *nĕphiyl,* which likely derives from the word *"napal".* Acknowledging that fact, we may still refer to them as the Nephilim, which should cause us no problem and should not become a source of confusion later. *Nephilim* is just the plural term of a *nĕphiyl* like *cherubim* is the plural form of *cherub.* The New International Version Bible refers to these hybrid beings as *"Nephilim"*, whereas the KJV (King James Version) refers to them as *"giants"*. There is another term, the *Rephaim*, which is not exactly a synonym – however, do see our footnote on the Rephaim for a bit more detail illustrating the confusion.[5] We will discuss Rephaim later.

To reiterate, even though we are citing the KJV, it should not be a problem to refer to these beings by using the word *Nephilim*. In our modern vernac-

---

[5] The definition from the *Bible Study Tools* testifies to the typical confusion: "Rephaim lofty men; giants, (Genesis 14:5; 2 Sam Genesis 21:16. Genesis 21:18 , marg. A.V., Rapha, marg. RSV, Raphah; Deuteronomy 3:13 , RSV; A.V., "giants"). The aborigines of Palestine, afterwards conquered and dispossessed by the Canaanite tribes, are classed under this general title. They were known to the Moabites as Emim, i.e., "fearful", (Deuteronomy 2:11), and to the Ammonites as Zamzummim. Some of them found refuge among the Philistines, and [still existed] in the days of David. We know nothing of their origin. **They were not necessarily connected with the "giants"** (RSV, "Nephilim") of Genesis 6:4." (Emphasis added) See http://www.biblestudytools.com/ dictionary/rephaim/. This definition fails to let the Biblical record speak plainly. Their existence originated from the angelic-human union, however hard to believe.

ular, the word *Nephilim* is the most common and recognizable word referring to the giant **offspring** of the fallen angels and human women. Thus, instead of merely referring to them with the more ambiguous term *giants*, we will refer to them as *Nephilim*. What isn't so obvious, but should be from the description we've just supplied: the Nephilim are **hybrids** of angels and humans. They are not fallen angels. Neither are the Nephilim humans with a minor amount of "alien DNA". They began with a 50/50 mix. (With each generation of course, the percentage of "angelic DNA" decreases mathematically depending upon the genetic make-up of the couple and whether Nephilim mated with other Nephilim or whether they were physically able to mate with other human women). Oftentimes, the word Nephilim is used interchangeably with the fallen angels. While **that is incorrect**, it is acceptable to use the term synonymously with the biblical term mentioned briefly above – *Rephaim*. However, keep in mind that neither word should be seen as a synonym for *fallen angels*.[6]

The Rephaim are mentioned many times in the Old Testament. We cite over two dozen uses of the term here: Genesis 15:20, Deuteronomy 2:11, 20, 3:11, 3:13; Joshua 12:4, 13:12, 15:8, 17:15, 18:16; and seven other references exist in 2 Samuel, five references in 1 Chronicles, and even one reference in Isaiah. Once you discover this strange fact about "hybrid breeding" in the biblical record, *you begin to see giants everywhere –* "Your eyes are opened" as it were. But, henceforth we'll just stick with the term *Nephilim*.

Another interesting point to share: the definition of the Hebrew word *napal* ("to fall") and can also mean "cast down, die, fall, fallen, inferior, be lost, perish, rot, throw down". The Nephilim, metaphorically speaking, were "fallen" too, being hybrids (neither fish nor fowl so to speak), and were lost

---

[6] There is some controversy about whether Nephilim actually refers to the angels or sons of God who fell, or refers to the offspring created from their illicit cohabitation with human women. Since *něphiyl* means "faller" according to Gesenius (Strong's H5303) on the surface it would seem we are talking about the angels and not the "giants". However, Numbers 13:33 uses the term Nephilim twice referring to the giants, i.e., the **offspring**, not the angels. *"And there we saw the giants,* [H5303] *the sons of Anak, which come of the giants:* [H5303] *and we were in our own sight as grasshoppers, and so we were in their sight."*

beings that perished *with no hope of salvation*. But Nephilim were not fallen from heaven as the fallen angels literally were, but were half-breeds instead of thoroughbreds – and were fallen in the other sense of the term; that is, *inferior, destined for death*, etc. The angels once had a high estate in Heaven, but disobeyed God and then fell from their high status (and apparently a sinless state) – perhaps even inheriting a "sin nature" just as we human beings acquired after our "fall".[7]

A COSMIC CONNECTION

On the other hand, the Nephilim, were lesser beings from their birth. They were always known as *just* Nephilim – not human, not angel – even from their moment of origin. Intelligent? Yes. Sentient? Yes. But not capable of being "tuned into God" or metaphorically, being on the same frequency with the LORD of Heaven. There could be no inner interconnection. One of the distinguishing features of being human is that "big something" that the Nephilim lacked. Because we possess a spirit, we have the ability to connect directly with God.

In other words, the inner connection between humanity and the LORD above is something special that *people* possess. It is quite possible that this is another factor that distinguishes human beings from the angels. Angels are in the presence of God, but we are not aware whether the LORD God dwells within them as He does us (He dwells within us IF we have been regenerated by the "second birth" – being born again as Jesus taught Nicodemus). We can be spiritually "invaded" by a spirit being. But how God's Spirit comes to dwell in us constitutes something quite different from other spirits coming into us. Such other spirits dominate, harass, and even possess us where we are no longer in control. The communion that Christians experience with Christ and the Father through the Holy Spirit encourages, comforts, and always remains peaceable. Yahweh "fellowships" with us.

---

[7] What is a sin nature? It is an "impulse or inclination to sin" instead of an impulse to behave in accordance to the will of God – see Romans chapters 6 and 7. Its purpose is to drive us to God for His solution to our sin problem – but that's another story.

Of relevance here, we should emphasize that *our connection is cosmic.* *"That which we have seen and heard declare we unto you, that ye also may have fellowship with us: and truly our fellowship is with the Father, and with his Son Jesus Christ."* (1 John 1:3) When thinking of what constitutes a Christian Cosmology, we must begin with this truth.

Granted, we don't often think about how central "having a relationship with Jesus Christ" through our spirits is *to the cosmology of the universe.* But it is perhaps *one of the most important of all aspects of the entire Cosmos.* And its implication is that at its very core, *the Cosmos is personal and moral, since Jehovah is personal and moral* and He created beings within whom He can dwell – but only on the basis of those creatures possessing moral righteousness (a righteousness that can only come through the redemption purchased for us through the death of Jesus Christ).

Figure 15 - THE APOCRYPHAL BOOK OF ENOCH

2 Corinthians 5:17 teaches this, *"Therefore if any man be in Christ, he is a new creature: old things are passed away; behold, all things are become new."* On this basis, the LORD God dwells within believers.

Returning to the fallen angels, we speculate that these fallen angels sought to beget offspring after their image. But God denied them this right. They could not beget new life like themselves without human women who had the capacity to conceive new life. Thus, the newly proud parents of the Nephilim – the fallen angels – soon learned that their children were unredeemable. At birth, before they had done anything in their bodies good or bad, the Nephilim had no hope of redemption. In a spiritual sense, they were dead from conception.

In the apocryphal *Book of Enoch*, it appears this realization created immense sorrow within the fallen angels. Additionally, we should note the Nephilim were horrifically vicious; one wonders whether they might have been made even more violent as they became fully aware they had zero chance for redemption and therefore, lacked any motivation to follow the moral demands of God. In a very real sense, having no opportunity for resurrection, the destruction of the Nephilim at the flood only hastened the inevitable death they faced. (We should mention at this point that most scholars through the ages have maintained that, upon their death, the bodiless spirits of the Nephilim became the ghosts and demons that inhabit the hidden dimensions of our universe and literally haunt human beings; especially when we seek "paranormal activity"). Interestingly, *rephaim*, another word for giants we mentioned earlier, *is the modern Hebrew term for ghosts.*

"No redemption for the Nephilim" seems like a very harsh judgment to us today. But we need to be aware what was at stake when these things happened and why this outcome was necessary.

When God created Adam, the very first man, He breathed into him the breath of life. Adam became a living soul (Genesis 2:7). Since then, every descendant of Adam has had this spiritual component in his or her make-up, a gifted life force often referred to as a "soul". Many would depict the soul as the *psyche* – equivalent to the mind and the emotions – what makes each of us who we are as individuals. But the Bible also conveys that humans have a "spirit" and the Judeo-Christian worldview distinguishes between the soul and the spirit (Hebrews 4:12, 1 Thessalonians 5:23). We believe the spirit is a faculty of human beings that enables us to encounter and experience the LORD God in a dimension different than we encounter other humans. Although this faculty needs regeneration to be fully alive, the faculty is passed down generation to generation as part of what makes us human. Clearly, this same capacity did not exist within the Nephilim. Knowing God "internally, within their spirits" was not a Nephilim capacity and therefore, not a dimension to their nature as it is ours.

The Nephilim were beings that were never meant to exist on this planet. Given the Hebrew definition of their name, we can see that God did not bless these creatures with the breath of spiritual life, such that they were able to commune with God through their spirits. They were not made in God's image. While angels, even fallen angels, are children of God (or sons of God), the Nephilim were in no sense children of God. They were children of the fallen angels.

FALLEN ANGELS

So after the fallen angels mated with human women, the Nephilim were born. From the descriptions given to us in the Bible, such as the accounts in Genesis and most famously in the story of Goliath, we learned that these Nephilim were giants, exceedingly strong, and barbaric. Not only were they extremely tall (Og King of Bashan had a bed at least 13-feet long), they had six fingers on each hand and six toes on each foot.

By committing this act of procreating with human women, the fallen angels were in direct violation of God's commandments. While we read about this encounter in the Genesis 6 account, there is a familiar verse in the New Testament that sheds additional light on how far "out of bounds" the angels had gone.

Speaking on the attributes of angels in Matthew 22:30 Jesus says, *"For in the resurrection they neither marry, nor are given in marriage, but are as the angels of God in heaven."* The LORD did not create angels to be bound together spiritually and physically in the same way a man and woman are on this earth. Likewise, angels were never meant to create offspring in their own image.

However, a common misconception exists regarding Jesus' statement. When it's cited, typically the last part of the verse *"of God in heaven"* is omitted. Sometimes translators amplify the verse implying that no angel has ever taken part in mating or procreating. That "improvement" upon what was said would be inferring something that the verse does not say. Jesus' statement in such cases has been taken out of context. This verse describes *"the angels of God in heaven"* but it does not necessarily depict angels who once resided on this Earth. The angels who took part in the unholy union with human women ceased functioning as angels of God. Once this occurred, apparently they could no longer reside or even access Heaven and appear before God. In stark contrast, angels still abiding in Heaven never committed this horrific trespass against God's law.

We refer to fallen angels as *fallen* because the Bible says "they left their first estate". We read about this in Jude 1:6-8...

> *And the angels which kept not their first estate, but left their own habitation, he [God] hath reserved in everlasting chains under darkness unto the judgment of the great day. Even as Sodom and Gomorrah, and the cities about them in like manner, giving themselves over to fornication, and going after strange flesh, are set forth for an example, suffering the vengeance of eternal fire. Likewise, also these filthy dreamers defile the flesh, despise dominion, and speak evil of dignities.*

We should point out several other truths that leap from this passage. First, through their sin of mating with humans (*"going after strange flesh"*), we see that fallen angels are compared to the actions of the people of Sodom and Gomorrah (and the surrounding cities), which resulted in God locking up these angels "in chains" (a supernatural restraint of some sort) until the moment known as the great "White Throne Judgment" when these angels along with unrepentant humans will suffer the *"vengeance of eternal fire"*.

A basic understanding of this passage in our English Bible supplies some interesting detail to be sure. But when looking at the original Greek, we gain a much keener appreciation for what was being stated.

For starters, we see the phrase, *"their first estate"*. The word *first* comes from the Greek word *arche,* meaning "a high rank", such as "to be first in position of political power". Then the word *estate comes* from the Greek word *peri,* which can mean a "locality". From this combination of terms, we conclude that the angels did not *keep their highest-ranking location.*

That is, Heaven comprised the highest-ranking location an angel could enjoy. Heaven stands as "elevated" in status as any place can possibly be. And yet, the angels gave this up. It was like forsaking the White House to live in "the Projects". Why is this so?

Perhaps they forfeited their birthright because they were jealous of the human capacity to procreate. Consequently, they decided the "grass was greener" on earth than it was in Heaven (one wonders if that might literally be true!) It's certainly hard to imagine for the earth-bound, who often dream of Heaven. But this may not only have been their only or even primary motivation.

Figure 16 - GATE OF FALLEN ANGELS, MOUNT HERMON

The next portion of the verse provides another insight, *"(they) left their (own) habitation..."* The word *habitation* comes from the Greek word *oiketerion* and means "a residence", such as "a family household". This may be saying that certain angels left their heavenly family and their heavenly home (their residence with God and the other angelic beings). It seems likely that these angels had taken part in the earlier rebellion –

that they were part of the satanic revolt that Revelation tells us resulted in "one-third of the angelic host" being cast down to earth.

As if that wasn't bad enough, a subset of this vast number of angels – the *Book of Enoch* tells us the subset numbered 200 (Enoch 6:5,8) – elected to come down upon Mount Hermon (note: originally the mountain was named Bashan), and then conspired to carry out their plan to "take wives for themselves from the daughters of men" having lusted after them. Their actions here on Earth resulted, ultimately, in God binding them in some sort of "heavy chains" and keeping them in gloomy darkness while also imposing the death sentence – being cast into and enduring eternally in the lake of fire.

Peter teaches, *"For ... God spared not the angels that sinned, but cast them down to hell, and delivered them into chains of darkness, to be reserved unto judgment..."* (2 Peter 2:4). But the meaning may simply reiterate that the angels inhabited one form of body and exchanged it for another form when falling to earth. If this is the meaning, then "habitation" refers to material bodies that functioned in space-time and may then be subject to weaknesses of the "flesh". Did the angels lust after the daughters of men after they took on an earthly body? If so, this explains to some extent how angels (no longer living in an angelic body) could "fornicate" with women as Jude declared.

OUR HEAVENLY HOME – AN IMMORTAL HOUSE OF FLESH

If we consider 2 Corinthians 5:2, we see that Paul speaks about being *"clothed upon with our house which is from heaven"*. In context, it becomes evident the *"house"* comprises a metaphor for a body. It is possible that the fallen angels not only forsook God and Heaven when they rebelled with Satan, but also forsook their heavenly spiritual bodies for a type of physical body they could use to operate within our physical dimension. If you saw the movie, *City of Angels* starring Nicholas Cage and Meg Ryan (featuring that great song *Iris* by the *Goo Goo Dolls*), you learned in the movie that the angel, played by Cage, had to "give up his wings" (and his immortality) to be with his love, played by Ryan. Did this end well? Well, we won't spoil it for you. But given the context of the reference, you can probably guess.

We can understand just how serious the angel's actions were given the nature of God's judgment upon them. This passage compares what they did to the reproachable sins of Sodom and Gomorrah. It says they were *"...giving themselves over to fornication, and going after strange flesh..."* The word *strange* comes from the Greek word *heteros* meaning "another, the other, or different" (we use the term in English as a prefix to denote the opposite of what is standard or accepted, such as *heterodox* compared to *orthodox*). This is saying that the fallen angels were engaging with humans in an act of procreation that was akin to what was considered fornication at Sodom and Gomorrah. They were "interfacing" with flesh that was different than their own type of "flesh", obviously in a forbidden way, not in keeping with the original design, plan, or purpose for which they were created by God.

There are several theories regarding how this could be possible; and it is not likely we will understand these things, for the scripture just doesn't elaborate. Nevertheless, the fallen angels possess some manner of physical body. When we pause to reflect on it, most of us recall a number of biblical passages that mention angels appearing as men and even having physical attributes. Jacob wrestled with an angel in Genesis 32:24 – probably, the pre-incarnate Jesus in angelic form (known as the "Angel of God"). This amazing wrestling match served through the centuries as the basis for many sermons seeking insight into the real meaning of this cosmic tussle. However, our point here is simply this: God often appears as an angel in the Old Testament. In fact, he appears sometimes as a light or fire (recall the burning bush) and as an angel (consider his discussion with Abraham over whether Sodom and Gomorrah should be spared). This bodily form of the "Pre-existent Christ" seems quite distinct than his human form after his birth on this earth, and as the glorified Christ witnessed in Revelation, chapter 1:14-15:

> *And in the midst of the seven candlesticks one like unto the Son of man, clothed with a garment down to the foot, and girt about the paps with a*

golden girdle. His head and his hairs were white like wool, as white as snow; and his eyes were as a flame of fire; And his feet like unto fine brass, as if they burned in a furnace; and his voice as the sound of many waters. And he had in his right hand seven stars: and out of his mouth went a sharp two-edged sword: and his countenance was as the sun shineth in his strength.

The nature of Jesus' immortal but physical body after His resurrection also comprises a different type of body from these other two instances above. After the resurrection (but before His ascension) Jesus could consume food but he could still pass through walls and move from place to place at will. (Isn't it interesting that after Jesus ate and went through the wall of the "upper room", the food he had swallowed went with him!")

The nature of this body stands as the prototype for our immortal body: *"Beloved, now are we the sons of God, and it doth not yet appear what we shall be: but we know that, when he shall appear, we shall be like him; for we shall see him as he is."* (1 John 3:2) And likewise, Paul teaches, *"For our conversation is in heaven; from whence also we look for the Saviour, the Lord Jesus Christ: Who shall change our vile body, that it may be fashioned like unto his glorious body, according to the working whereby he is able even to subdue all things unto himself."* (Philippians 3:20-21)

In comprehending the Cosmos and grasping what is important to the ultimate "nature" of creation, a corporeal estate – a bodily existence, fashioned in the manner God intended – comprises one of its most important elements. **Existence is not meant to be wispy or ethereal as the Greeks conceived. It is to be conscious, tangible, and sensual in a pure and perfect sense.** Paul cites Isaiah 64:4 (and annotates it a bit): *"But as it is written, 'Eye hath not seen, nor ear heard, neither have entered into the heart of man, the things which God hath prepared for them that love him.'"* (1 Corinthians 2:9) [8] Sensations and emotions will be part of what makes us *us* in the age to come. Jesus even makes the point that he will not have wine in heaven until he is with us, *"Verily I say unto you, I will drink no more of*

---

[8] *"For since the beginning of the world men have not heard, nor perceived by the ear, neither hath the eye seen, O God, beside thee, what he hath prepared for him that waiteth for him."* (Isaiah 64:4)

the fruit of the vine, until that day that I drink it new in the kingdom of God." (Mark 14:25) For those of us that enjoy the pleasure of good food and "the fruit of the vine" (in moderation of course), there is good news: We will enjoy them for eternity.

In much the same way, we see angels in the Old Testament able to consume food and drink. We recall the three angels sitting down and eating with Abraham after Sarah made them lunch (Genesis 18:1-15). Although the circumstances are different from one another, we still see that angels possess some form of physical body – it's much stronger, it's immortal, and it can move from place to place. It can receive food and drink. However, when it comes to the most incredible account, in which they procreated with human beings, we must admit we aren't given the details. It isn't likely we will understand how fallen angels could procreate with human women (or why they "lusted after the daughters of men") this side of the Second Advent. Was the embryo formed through cellular mitosis? Was some sort of advanced genetic technology involved? We will consider this topic in the next chapter. But the result of a physical union is plainly stated in the Old Testament, confirmed in the New Testament, and made more explicit in apocryphal books, like the *Book of Enoch*, the *Book of Jasher*, and several others. Angels and humans mate in the corporeal, physical sense – and hybrid offspring result. **And this has big implications for the Cosmos.**

Although it stretches our minds, when we think about what we comprise *in corporeal form*, there exists reason to accept that this can all be true. We humans are *tripartite* beings (composed of three parts), body, soul, and spirit.[9] The ethereal soul and spirit has been placed into a physical structure so we may operate within the physical reality. Our physical bodies are a "sense vehicle" for our souls and spirits to exist within this physical realm, a reality we experience and encounter through sights, sounds, smells, tastes, touches; and also as inner sensations of all kinds – aches, pains, excitement, longings of hope, intense pleasure and so on.

---

[9] *"And the very God of peace sanctify you wholly; and I pray God your whole spirit and soul and body be preserved blameless unto the coming of our Lord Jesus Christ."* (1 Thessalonians 5:23)

It would seem reasonable that another type of spiritual being, like an angel, might seek help from our "human nature" (literally) to operate effectively in our physical dimension.

We presume all angels are masculine – which appears to be the Bible's plain teaching – Michael isn't Michelle and Gabriel isn't Gabriela. However, one could assume instead that they are androgynous beings without gender, but which still require a corporeal body to interact "normally" with the physical realm. This body would be different than ours, as we were created originally a *"little lower than the angels"* (without some of their capabilities – Hebrews 2:7). We know from biblical accounts, angelic bodies are more powerful and able to do things that the human body can't do. The substance that composes an angelic body might be different than human flesh. Therefore, the term "flesh" possibly constitutes a mere metaphor for this "material". We really do not know if "flesh" remains an appropriate term to describe an angel's body because it's clearly distinct from what we humans possess. However, we certainly should feel free to employ the metaphor and refer to angelic "body material" as flesh.

We should recall also that Paul teaches us that there are different kinds of flesh and different kinds of bodies, some mortal and some immortal. Paul says one day we will have a new body, i.e., inhabit a "new house". Our current estate is corporeal but subject to decay ("depreciation" big time) and is but an illustration of what it will truly be. When it takes on immortality, our estate will be purely real. You might describe our situation as "real estate" (forgive the pun, but we hope it makes the point memorable).

> *For which cause we faint not; but though our outward man perish, yet the inward man is renewed day by day. For our light affliction, which is but for a moment, worketh for us a far more exceeding and **eternal weight of glory**. While we look not at the things which are seen, but at the things which are not seen: for the things which are seen are temporal; but the things which are not seen are eternal.* (2 Corinthians 4:16-18)

And of course the famous passage on resurrection: *"For this corruptible must put on incorruption, and this mortal must put on immortality."* (1 Corinthians 15:53). For now, however, we must hang out in the *habitation*

NOAH

Returning again to Genesis: Genesis 6:9 states, *"These are the generations of Noah: Noah was a just man and perfect in his generations, and Noah walked with God."* What interests us here: there are two different Hebrew words used for our English word *generations*. The first time it is used is the Hebrew word *toleda* meaning "family descent". When the second Hebrew word *dor* is used, it means "a revolution of time", or "an age". The word *perfect* comes from the Hebrew word *tamim* and means, "without blemish, complete, and undefiled". Going back to the passage in the Book of Jude, the word for *defiled* also means "corrupt", as in "corrupted flesh".

Now Genesis 6:9 introduces Noah's heritage. And it conveys that Noah was *"a just man, and uncorrupted in all his revolutions of time"*. If only Noah and Noah alone was perfect, by deduction, we must conclude the rest of the world was in some sense defiled. This is what the Book of Jude implies – that fallen angels tainted human flesh by mating with the daughters of men, over *many* generations. Genesis 6:9 conveys that back in that time Noah and his descendants were the only people in the world that were completely **uncorrupted**.

In other words, the rest of the world was debased by hybridization, aka the Nephilim. However, Noah's bloodline, beginning with Adam, from generation to generation, was completely untainted by any angelic material. Its "perfection" testified to its purity – meaning no one in the bloodline of Noah back to Adam ever mated with a fallen angel or a descendant of a Nephilim. Why was this important? Scholars tell us that Satan's master plan was *to eliminate the possibility of humanity's redemption by corrupting humanity's genome.* More specifically, Satan's plan was *to corrupt the bloodline of the*

Savior. Noah's **perfect** bloodline, untainted by the Nephilim, was the singular bloodline emerging from the antediluvian world, through which Jesus Christ would be born.

J.R. Church's last book before his departure to be with our LORD wrote *Enoch: The First Book Ever Written.* Church commented concerning the issue of angels corrupting the human race:

> [Chapter 6] reveal(s) one of the main purposes for [Enoch] writing this ancient book. We are told that fallen angels, cohabiting with human women were the specific reason for initiating the catastrophe of Noah's Flood. God's purpose for the redemption of mankind could have been thwarted by a corruption of the genetic code.[10]

Figure 17 - LOT AND HIS DAUGHTERS BY LUCAS VAN LEYDEN (1520)

That explains why God saved only Noah and his family from the destruction of the flood. It also explains why God wanted the tainted flesh of the Nephilim eradicated from the earth. Those commandments to Joshua and the Israelites to "leave no man, woman, or child alive" when engaging in battles to conquer the Promised Land were given by God expressly

---

[10] J.R. Church, *Enoch: The First Book Ever Written*, Oklahoma City: Prophecy Publications, 2015, p. 35.

for this reason. Pause to consider this. It's no small thing. **It is indeed crucial to understand the Cosmos.**

To take this one step further: since Jesus was born without blemish, he was born through a "perfect" 100% human bloodline. Satan's attempted corruption of the entire human population was to deprive God of redeeming humanity by sending a Savior. Had God not intervened and judged the Earth through the Great Flood thwarting Satan's plan, humankind would have no future and all prior believers before Jesus' day, would be lost as would we, without hope. The flood eradicated the corruption Satan had wrought upon the earth through the angelic incursion *prior to Noah.* But wait – was that the end of the matter? We learn the answer is "No". The drama intensifies.

THE SECOND INCURSION – FIVE DIFFERENT THEORIES

The Bible indicates that angelic incursions tainting the bloodline of humanity were not entirely eliminated by the destruction of all life on earth (save Noah and his family) through the Great Flood. In other words, the bloodline of the future Messiah remained under attack after Noah's Ark came to rest and humanity went about the process of replenishing the earth. We deduce this because there are so many accounts of giants *after the flood.*

One such account can be found in Numbers 13:33, which reads, *"And there we saw the giants [Nephilim], the sons of Anak, which come of the giants [Nephilim]: and we were in our own sight as grasshoppers, and so we were in their sight."* Another example can be found in Amos 2:9, which reads, *"Yet destroyed I the **Amorite** before them, whose height was like the height of the cedars, and he was strong as the oaks; yet I destroyed his fruit from above, and his roots from beneath."* These passages, as well as numerous other passages in the Bible (one recalls the story of David and Goliath), tell us that after the flood, giants were once again present on the earth.

There are five distinct theories which attempt to explain the origin of the second influx of Nephilim. Unfortunately, the Bible isn't explicit concerning the origin of the second incursion of the Nephilim, so we can only speculate as to why Nephilim still existed after the Flood.

Theory number one: The fallen angels mated with human women again, after the flood, meaning there was a second and separate incursion of fallen angels. If true, this second band of angels didn't learn the lesson other angels already had – that Yahweh didn't take kindly to messing with the human genome. This is certainly possible since it had already happened once before. But it suggests that God's solution – the Great Flood – failed to solve the problem.

Theory number two: Fallen angels didn't cause the reappearance of the Nephilim; the giants reappeared due to corrupted human genes "coming aboard" the Ark. From our prior study we reiterate two things: first, Noah was *"perfect in all his generations"* (Genesis 6:9); and secondly, all other *humans were corrupted* (Genesis 6:12). In this second theory, it is supposed that Noah married his sister (her name is never mentioned in the Bible although she is referenced five times). If so, their three sons would be of pure human "flesh". However, since all other flesh was corrupted, the wives of Noah's sons would not have been, unless, once again, the sons married Noah's daughters not mentioned in the account (while possible, this does not seem to have been the case). One or more of the wives carried Nephilim-tainted genes.

There is some intriguing genealogical evidence uncovered through careful biblical sleuthing that may validate this theory. Noah's son Ham had four sons, Cush, Mizraim, Put and **Canaan**. Apparently, through Canaan's eleven (11) sons that settled the land of Canaan, by the time of the Exodus centuries later, giants were everywhere throughout Canaan. As mentioned earlier, this is apparent when the Israelites returned to take the Promised Land. Author and researcher Rob Skiba has done some excellent work in this respect and argues a strong case that the Bible's historical record best supports this theory.[11] Others who have studied the subject disagree, such as L.A. Marzulli, favoring the first theory of "multiple incursions" before and after the Flood.

Theory number three: Survivors after the flood could have acquired some technological means to manipulate genetics. This seems highly unlikely, but angelic beings had a propensity to share technical know-how with ancient humanity (this is elaborated on in the *Book of Enoch*). So perhaps angels appeared once more and held classes to pervert humanity's approach to procreation and stop sticking to Yahweh's rules of "reproducing after their kind." There are different variations of this theory, but all have to do with the bodies of the dead Nephilim after the flood. No doubt, the ground of the post-flood world could have been covered with the bodies of drowned Nephilim. If the inhabitants of the world had knowledge of gene manipulation, they could have manipulated the Nephilim genes into their own. Think Jurassic Park – dinosaur DNA that was the source for regeneration of T-Rex and his friends. This type of genetic work is becoming commonplace today.

Theory number four: A **fourth theory** follows along the same lines and it's equally breathtaking. Ham's son **Cush** had six sons, five of which became the heads of clans populating northern Africa. However, the sixth son was Nimrod. Highly regarded author, researcher, and publisher Tom Horn speculates Nimrod may have become a Nephilim through some undisclosed process, introducing another relevant and highly intriguing term, *gibborim*, "a mighty one". Says Horn,

> The story of Nimrod in the book of Genesis may illustrate how this could happen through genetic engineering or a retrovirus of demonic design that integrates with a host's genome and rewrites the living specimen's DNA, thus making it a "fit extension" or host for infection by the entity. Note what Genesis 10:8 says about Nimrod: *"And Cush begat Nimrod: he began to be a mighty one in the earth."*

> Three sections in this unprecedented verse indicate something very peculiar happened to Nimrod. First, note where the text says, "he began to be."

---

[11] Rob Skiba, Archon Invasion: *The Rise, Fall, and Return of the Nephilim.*

In Hebrew, this is *chalal*, which means "to become profaned, defiled, polluted, or desecrated ritually, sexually or genetically." Second, this verse tells us exactly what Nimrod began to be as he changed genetically – "a mighty one" (*gibbowr, gibborim*), one of the offspring of Nephilim. As Annette Yoshiko Reed says in the Cambridge University book, *Fallen Angels and the History of Judaism and Christianity*, "The Nephilim of Genesis 6:4 are always…grouped together with the gibborim as the progeny of the Watchers and human women."[12] And the third part of this text says the change to Nimrod started while he was on "earth." Therefore, in modern language, this text could accurately be translated to say: "And Nimrod began to change genetically, becoming a **gibborim**, the offspring of watchers on earth." [13]

Theory number five: Some hybrids didn't drown in the flood – they found a way to survive. Some flood myths from ancient peoples teach this. Others have speculated this could be so, but it contradicts what the Bible teaches: *that all flesh was destroyed.* If there were survivors, it would appear to indicate *the testimony of the Bible is in some way not true.* Still, if for no other reason than the fact some flood myths from other cultures state this to be so, we might not dismiss this possibility out of hand. But it is not the favorite theory most evangelicals advocate for obvious reasons.

To recap: from the biblical account of Genesis 6:1-4, there was an original incursion of Nephilim before The Great Flood; a flood that by most accounts utterly destroyed them. We know there was a second incursion sometime after the flood since giants appear again and have populated Canaan during the 400 years the Israelites were slaves in Egypt and sojourned in the Wilderness. However, based solely on the Biblical record, we cannot be certain why the Nephilim reappeared.

The story of the Promised Land's conquest by Joshua, Caleb and the other Israelites includes this element as an essential plot line. Giants were the primary obstacles the Israelites had to overcome to win back the land promised to Abraham. To reinforce the "size of the problem" (pun intended),

---

[12] Annette Yoshiko Reed, *Fallen Angels and the History of Judaism and Christianity: The Reception of Enochic* Literature (Cambridge, 2005), p.214

[13] See http://www.newswithviews.com/Horn/thomas155.htm#_ftn1.

from examining Old Testament accounts there may have been over 30 different tribes of giants.[14] The Amalekites, Zamzummims, and Zumins were three such tribal names often linked or expressly stated in the Bible as giants. Numbers 13:30-33 is the prime example, confirming that the giants were inhabitants of Canaan. Joshua and Caleb along with ten others "spied out" the land. Despite witnessing the giants of Canaan up close, Caleb and Joshua still recommended the Israelites go forth and take the land:

> *And Caleb stilled the people before Moses, and said, "Let us go up at once, and possess it; for we are well able to overcome it." But the men that went up with him said, "We be not able to go up against the people; for they are stronger than we." And they brought up an evil report of the land which they had searched unto the children of Israel, saying, "The land, through which we have gone to search it, is a land that eateth up the inhabitants thereof; and all the people that we saw in it are men of a great stature. And there we saw the giants, the sons of Anak, which come of the giants: and we were in our own sight as grasshoppers, and so we were in their sight.*

But the Hebrews did not proceed forth. They rejected "the minority report" of Joshua and Caleb, and God effectively cursed that generation, indicating that the entire generation of adults who went along with "the evil report" would not see the Promised Land. So 40 years of wandering in the wilderness resulted.

Meanwhile, the giants continued in the land of Canaan. The Bible says that *"the iniquity of the Amorites was not yet full."* (Genesis 15:16) Joshua and Caleb entered the Promised Land after the years in the wilderness; Moses saw the Land from Mount Nebo but did not enter the land; but the rest of the Israelites "of age", from 40 years before, perished in the wilderness.

Some have speculated that after the Hebrews came out of the wilderness and established "the LORD was with them" through several early victories,

---

[14] Tribal names in the Bible for giants may include: Amalekites, Amorites, Anakims, Ashdothites, Aviums, Avites, Canaanites, Caphtorims, Ekronites, Emins, Eshkalonites, Gazathites, Geshurites, Gibeonites, Giblites, Girgashites, Gittites, Hittites, Hivites, Horims, Horites, Jebusites, Kadmonites, Kenites, Kenizzites, Maachathites, Manassites, Perizzites, Sidonians, Zamzummims, Zebusites, and Zumins.

the giants in Canaan began to flee. This watershed moment came on the heels of the Battle of Jericho. Once the inhabitants throughout the land (mostly giants!) heard how the LORD destroyed the city of Jericho (when "the walls came tumbling down"), such accounts of the LORD being with Moses, Joshua, and Caleb became "headline news" within Canaan. (See Joshua 2 – 6, 9-11). Many battles were fought and the giants were defeated again and again. How complete was the conquest? *"There was none of the Anakims left in the land of the children of Israel: only in Gaza, in Gath, and in Ashdod, there remained."* (Joshua 11:22) And of course, Goliath and his brothers (also giants) lived in the town of Gath, when two centuries later, he would be defeated and killed by the young David.

In fact, it is possible that the giants fled to the New World (which has been argued by L.A. Marzulli and separately by Woodward). The frequent record of giants discovered in the New World has given rise to additional speculation that many reports printed in newspapers during the nineteenth century were intentionally downplayed at the turn of the century (so says L.A. Marzulli, Steve Quayle, and Jim Vieira). These researchers /authors assert the Smithsonian Institute made it standard practice to go from place to place where the bones of giants were discovered (giants typically ranging from 7 to 9 feet tall) to secret away the bones and hide them in an "Indiana Jones-like warehouse" perhaps in the outskirts of Washington D.C. Whether the Smithsonian in fact followed this practice may never be confirmed. Nevertheless, first-hand accounts of "the Feds" absconding with the remains of giants have been reported and discussed frequently in alternative media.

NEPHILIM AND THE LAST DAYS

But is it possible a third incursion of angels mating with the "daughters of Adam" will occur and literally give birth to hybrids? And, if so, could we be witnessing this incursion in our day?

In Luke 17:26-30, Jesus preached that the last days would be like the days of Noah and Lot. Therefore, He would not establish His Kingdom until after conditions in the world become similar to the "life and times of Noah". As for the days of Noah, Noah's contemporaries were not expecting the flood and couldn't appreciate the fact that someone would build a massive ship so far inland. Furthermore, for those not tuned into the signs of the last days nor spiritually prepared for their occurrence as with the scoffers who ignored Noah's preaching, by the time Christ does return their chance to repent will be passed – time will have run out.

While Jesus offered a very clear prediction of the apocalypse in this passage, we believe also hidden within is the truth of the Nephilim. Of all of God's unexpected judgments Jesus could have chosen, He selected Noah and Lot as His examples. We know that Noah's days were unique in no small part because of the presence of the Nephilim. As mentioned earlier, from 2 Peter 2:4-7, *"For if God spared not the angels that sinned, but cast them down to hell, and delivered them into chains of darkness, to be reserved unto judgment; And spared not the old world, but saved Noah the eighth person, a preacher of righteousness, bringing in the flood upon the world of the ungodly; And turning the cities of Sodom and Gomorrah into ashes condemned them with an overthrow, making them an ensample unto those that after should live ungodly; and delivered just Lot, vexed with the filthy conversation of the wicked."*

It is no coincidence that Peter's statement should be paired with the notable passage in Jude 1:6-8, the passage that speaks of the patriarch Abraham and his nephew Lot. Both passages in Peter and in Jude begin by mentioning *the angels that sinned.* The traditions of angel and human hybrids were common Mesopotamian myths. The Jews of Jesus day would have been quite familiar with these myths and well aware of the *Book of Enoch.* As we pointed out earlier, Jude even cites it directly. Additionally, we believe Jesus referenced the days of Noah and Lot to emphasize that "history would repeat itself" in the end days. The Nephilim and their progenitors, the fallen angels, were the featured stars of these passages just as they were in the Mesopotamian myths. The point should not go unnoticed however, that in the account of the two angels who visited Lot to rescue him and his family, the inhabitants of Sodom sought to engage in sexual relations with Lot's two special guests. Did the sex-crazed Sodomites know these out-of-towners were angels? Perhaps. If so, it reinforces the notion that angels could have been sexual creatures or at least that the people of Sodom wanted to reenact the mating of gods with humanity.

No doubt angels in human form possessed a comely appearance and attracted attention. Regardless, the fact remains Lot's guests *were* angels, they had been noticed by the Sodomites, and the people in Sodom nearly rioted trying to get the angels to come out so they could have their way with them. Instead, Lot tried to offer his virgin daughters to the crowd. (Lot knew he was entertaining Abraham's envoys – at the very least – and it seems likely he realized they were much more than mere men). The angels, however, would not allow Lot's negotiation to continue… not on their watch.

They opened the door, grabbed Lot by the arm and pulled him back into the house. Then they struck the people outside the door blind. Early the following morning, no doubt before sunrise, they escaped with Lot, his wife, and Lot's daughters. Lot's would-be son-in-laws scoffed at the thought of leaving Sodom (apparently betrothed to Lot's daughters while their marriages were not yet consummated). Consequently, they paid the price for their indifference. Like everyone else in the city, they were consumed in the judgment of fire and brimstone that fell later that day destroying all the people in the area.

WILL HYBRID BEINGS APPEAR AGAIN IN OUR DAY?

Here is yet another example of why it's so important to take other scriptures into account when examining a single Biblical passage. There is a possibility that during the last days, which we believe stare us in the face right now, fallen angels will once again engage in unholy union with mortal women to produce "hybrid humans" like the Nephilim, potentially through laboratory-like genetic manipulation, but also potentially through forbidden sexual encounters. **The population "at large" has been prepared to believe such things aren't that extraordinary**. There exists a whole genre of literature, television programs, and movies devoted to alien abduction, the production of alien/human hybrids and even web sites devoted to help parents who claim to have hybrid children! Additionally, there are ministries devoted to counseling humans (perhaps they aren't fully human) who claim to be hybrids. And among these counselors, debates have occurred and literature has been written concerning whether or not these hybrid entities have souls and therefore, can be saved!

Text from Daniel 2:40-43 challenges us with something else to consider in the context of hybrid beings. This arises from the interpretation Daniel provides King Nebuchadnezzar regarding his famous dream of the great colossus with the head of gold. Daniel informs the king that the magnificent statue represents four kingdoms of which Nebuchadnezzar is the first king (the head of gold). Kingdoms of bronze and silver will follow. Then Daniel shares the following regarding the fourth kingdom:

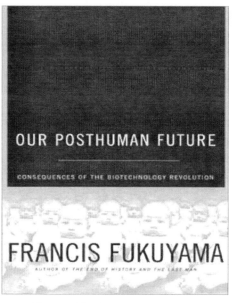

Figure 18 - *OUR POST-HUMAN FUTURE,* **FRANCIS FUKUYAMA (2002)**

And the fourth kingdom shall be strong as iron: forasmuch as iron breaketh in pieces and subdueth all things: and as iron that breaketh all these, shall it break in pieces and bruise. And whereas thou sawest the feet and toes, part of potters' clay, and part of iron, the kingdom shall be divided; but there shall be in it of the strength of the iron, forasmuch as thou sawest the iron mixed with miry clay. And as the toes of the feet were part of iron, and part of clay, so the kingdom shall be partly strong, and partly broken. And whereas thou sawest iron mixed with miry clay, they shall mingle themselves with the seed of men: but **they shall not cleave one to another, even as iron is not mixed with clay.**

Different interpretations exist for this stunning passage, specifically Daniel 2:43, based upon various premises about what lay at the heart of Nebuchadnezzar's dream.

Are we dealing merely with kingdoms? Are we concerned instead with the peoples that make up the populations of these kingdoms? Or is there something sinister about this final kingdom? Could the language in the last portion of this passage infer that the people of this fourth kingdom possess a hybrid nature? Have they been transformed into "trans-humans"? Are we in this fourth kingdom period now? Are we, as political philosopher Francis Fukuyama predicts, verging on a post-human future, the consequences of the biotechnology revolution? [15]

The last kingdom described by Daniel, has feet made of iron mixed with clay. The conventional view says that the image has unsure footing because the last empire does not rely upon authoritarian rule – it is based on a democratic republic. Others respond that such an interpretation comprises one very small part of *exegesis* and one very large part of *eisegesis*. Nevertheless, perhaps the most interesting portion of the last sentence lies with the phrase, *"they shall mingle themselves with the seed of men"*. While Daniel does not directly identify who *"they"* are, *they* must be different than the "seed of men". Certainly, humanity constitutes the *"seed of men"* as we have been procreated from the seed of our biological fathers. From the discussion in this chapter, we know that the only other entities that are not human but have had experience *"mingling themselves with the seed of men"* are the fallen angels creating the Nephilim. We believe it is not unreasonable to see in this passage a prophecy asserting that fallen angels will create (or are now in the process of already creating) a whole new generation of hybrid offspring in a last-gasp plan led by Satan himself, to corrupt the human race. Indeed, it has been commonplace nowadays to assume the "Mark of the Beast" may include some form of RNA retrovirus that will rewrite the human genome. The "Mark" will promise protection from disease and bring humanity to the cusp of immortality.

However fantastic or incredible this seems, we should not reject the possibility that the Bible through Daniel's interpretation of Nebuchadnez-

---

[15] Francis Fukuyama, *Our Posthuman Future: Consequences of the Biotechnology* Revolution, New York: Picador, 2002.

zar's dream, provides a subtle hint that the angelic-human hybrid phenomenon is underway once again especially if we are living in "the last days" – the days of the fourth empire.

CONCLUSION

Earlier, we mentioned in passing that the issue of alien abduction and the production of human/alien hybrids frequently occupies plots in television shows and movies. Documentaries (perhaps better labeled "mockumentaries"[16]) on the History Channel and Discovery TV, love to discuss UFOs and the possibility that alien races visited earth and provided genetic material to modify and enhance humanity in aeons past. This notion has generated a high level of public interest. We are now highly conditioned to judge this once crazy possibility as an established fact. Furthermore, as just mentioned regarding the "promise of the Mark of the Beast", humanity could easily be coaxed into believing that further modification of our human genome could eliminate disease, increase our mental faculties, and engineer us into "demigods" with superman features (enhanced vision, perception, speed, and strength). After all, who wouldn't want to be a superhero?

On the other hand, there are still many intellectuals and skeptics who scoff at such an idea. But before you join with the skeptics and dismiss these possibilities, we would strongly encourage you to consider the vast amount of evidence available to review. Research the issue before you make up your mind. Recall the wisdom of Proverbs 18:13, *"He that answereth a matter before he heareth it, it is folly and shame unto him."* Consider all the newly uncovered evidence for these things including the amount of science devoted to genetic engineering and enhancement, medicine based upon genetic splicing and recombination, and even building super soldiers. If it remains too hard to believe at this time, file it away for

---

[16] Wikipedia defines *mockumentary* as follows: "Mockumentary or mock documentary is a genre of film and television, a parody presented as a documentary recording real life. It is a comedy although in its presentation, it "keeps a straight face". Movies like *Spinal Tap*, *A Mighty Wind*, and *Waiting for Guffman*, by Christopher Guest have had a major impact on film. A TV Series like "The Office" is presented in the form of a mockumentary.

future reference. This discussion may later prove crucial to prepare you to accept such strange things in light of the most peculiar occurrences you may encounter in the days ahead, events you can grasp and reconcile to your worldview in no other way. We believe the "brave new world" of Huxley has become the world in which we live now.

To offer a simplified set of points to recap our discussion:

- The Bible teaches angels and humans once mated in ancient times creating hybrid offspring called Nephilim. These beings had unusual attributes including exceptional height and strength. They were labeled in the Old Testament by another generic name, the Rephaim. Many tribes of Nephilim lived in Canaan until after the Exodus from Egypt, when the Hebrews conquered the Promised Land and killed or drove out almost all of these giants.

- The New Testament writers, Jude and Peter, reference the non-canonical Book of Enoch thereby acknowledging and endorsing the concept of angelic/human interaction while indicating that such actions were reprehensible to God. The Great Flood of Noah can be explained as judgment and a means by which Jehovah could cleanse the earth of life corrupted by angelic incursion.

- Angels appear to have the capability to create beings "in their own image" only if they procreate with human women; but their offspring are not eligible to be redeemed by the blood of the cross and are sentenced to death and eternal punishment. There are some who do counseling today with individuals who claim to be hybrids, the offspring resulting from supposed "alien abductions" and impregnation by these aliens.

- Biblical scholars, often regard the bodiless spirits of the Nephilim, as the origin of demons.

- In today's world, it's generally agreed by scholars and writers on these subjects that the so-called aliens are really fallen angels (or bodiless spirits of the Nephilim), with an ability to manifest in our space-time realm with a physical body. We don't know all the details, but this notion seems readily reconcilable with Biblical accounts.

- In today's scientific world, the effort to improve upon the nature of the human race is called Transhumanism. The notion that animals can be given advanced traits that are unnatural to their species through genetic modification has become science fact, not fiction. We appear to be on the cusp of seeing this science become an accepted means to modify the human genome to "improve" humanity. The consequences could be disastrous. *"As in the days of Noah, so shall it be in the days of the coming of the Son of Man."* (Matthew 24:37) (This is discussed later in this book).

- The Prophet Daniel hints that the last kingdom may be comprised of individuals whose seed has been mingled with non-human seed. Hybrid creatures may already be among us today. Many sources, both Christian and non-Christian claim that this is so.

In closing, we should also reiterate that there exist today ministries and counseling services (who take the phenomenon at face value) addressing the needs of persons that believe they are hybrid or have had hybrid children. Several ministries exist that deal with human women who have been the subject of "interfacing" with angelic beings, creating hybrid children. These women are known as "hybrid mothers" and allege that the leaders of this fourth and final empire are hybrids "begotten" from the mating of angels with women of royal blood. In fact, one such ministry is built around a community of such persons who have provided accounts of their offspring who are "waiting in the wings" to step in as world leaders; specifically, the Antichrist and the False Prophet. [17]

But the story doesn't stop there. The ability for humanity to rediscover its capacity to cross over into other dimensions with enhanced skills like the "sons of God" will be taken up next. *Revising reality,* in accordance with the recent teachings and research of leading Christian thinkers today, mandates we explore the meaning of "ultra-dimensionality". A major portion of this book, necessarily, will also be devoted to this discussion.

---

[17] See the final two chapters of Woodward's book, *Power Quest, Book Two: The Ascendancy of Antichrist in America*, Faith-Happens, 2012) for a discussion on this community and its relationship to the Nazi infiltration into America, including the prominent role played by Josef Mengele in testimony more sinister than the fictional story, *The Boys from Brazil.*

DO ANGELS HAVE DNA?
Josh Peck

ANGELIC GIGANTISM

WITH THE BASICS REGARDING THE NEPHILIM COVERED IN THE LAST CHAPTER, YOU THE READER AND I, JOSH, CAN NOW EXPLORE SOME ADVANCED TOPICS. FOR EXAMPLE, THE QUESTION of size naturally comes up when researching Nephilim; that is: "Just how tall were these creatures?" Yet to me, the more interesting question and one rarely asked is, "**Why** were they so big?" For the scientifically minded, we deduce there must have been a physical, perhaps genetic reason for their enormous stature. Our discussion here links to the issue of "other dimensions" *which constitutes another key premise of the Cosmos.* That is, **a biblical Cosmology must emphasize this aspect of reality.**

On the surface, the answer would seem obvious. If they are part angel, it is no surprise that they would be giants. After all, angels were always considered very powerful beings. Still, diving in deeper we wonder what is it exactly that leads us to this conclusion. If we are to assume the Nephilim surpass the physical size of human beings (transcending eight feet tall – at least) and this was due to their angelic progenitors, we would no doubt assume that angels must also be gargantuan in physical size. Consider these passages from the Bible concerning angels:

> *And the LORD appeared unto him in the plains of Mamre: and he sat in the tent door in the heat of the day; and he lift up his eyes and looked, and, lo, three men stood by him: and when he saw them, he ran to meet them from the tent door, and bowed himself toward the ground (Genesis 18:1-2)*

> *And there came two angels to Sodom at even; and Lot sat in the gate of Sodom: and Lot seeing them rose up to meet them; and he bowed himself with his face toward the ground;*
>
> *And they called unto Lot, and said unto him, "Where are the men which came in to thee this night? Bring them out unto us, that we may know them." (Genesis 19:1, 5)*
>
> *Be not forgetful to entertain strangers: for thereby some have entertained angels unawares. (Hebrews 13:2)*

Genesis chapters 18 and 19 tell us the angels appeared as standard size men to those who witnessed them. There is no mention that they were bigger than the average man of that day. Hebrews 13:2 even tells us angels physically appear so normal, that people have entertained them without knowing they were angels.

An interesting side-note to this: there is no distinction made between heavenly and fallen angels; it only states "angels" without distinguishing the good guys from the bad guys. Both benevolent and malevolent angels appear typically human.

The biblical record does not indicate that a larger physical size is an attribute of angels. Why then would the hybrid offspring of fallen angels be so oversized? One possible answer would seem to have something to do with what Jude relates in his brief epistle. He infers the rebellious angels lost something intrinsic to their nature when they left their standard angelic but corporeal state.

> *And the angels which kept not their first estate, but left their own habitation, he hath reserved in everlasting chains under darkness unto the judgment of the great day. (Jude 1:6)*

As stated earlier, an interpretation could be set forth that argues this *first estate* or *habitation* refers to the angelic body and not their heavenly housing. However, this still brings up an important question. If the fallen angels truly traded in their extradimensional angelic body for a completely three-dimensional physical form, would that apply to genetics as well? That is, if there was nothing angelic anymore about their body, would their nature not also be transformed, going beyond their outward appearance? Wouldn't their cellular makeup, physiology, and genetics mechanisms be altered too? Would angels have DNA and would that

DNA be different than the building blocks of their corporeal nature when they lived in their "heavenly habitation" – their "first estate"?

To be a bit more direct, if we buy that there are angels at all, why wouldn't their corporeal nature include some manner of DNA?

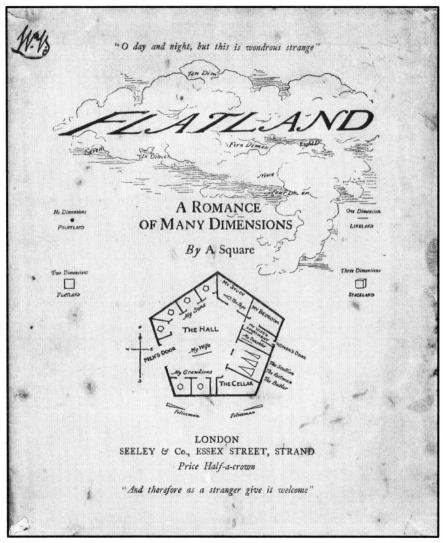

Figure 19 – FLATLAND BY EDWIN ABBOTT, COPY OF FIRST EDITION (1884)

I do not believe it would be accurate to say that the angels who fell from heaven traded their heavenly angelic bodies (their "first estate") for human bodies – not completely at least. I do believe, however, that the text suggests strongly a type of downgrade occurred, that amounted to more than going from an automatic to a manual transmission (so to speak). This is especially so when considering the meanings of the words the original authors employed. Having made this point, however, I believe one aspect of their angelic physiology remained: they retained *extradimensionality*.

SUPER-PHYSICALITY AND EXTRA DIMENSIONS

As I show in my books *Quantum Creation* and *Cherubim Chariots*, angels are true extradimensional beings. This applies both to heavenly members of the angelic realm in good standing *and* fallen angels who do not necessarily (may or may not) retain access to the throne of God in heaven. Instead of thinking of angels as ethereal, wispy entities that are made from cotton candy-like elements as our popular culture seems to think, it would be much more probable (and our thinking far more sophisticated) to think of angels as being *super-physical*. Angels possess *a superior* physicality to humankind with a more complex physiology based upon a higher order body chemistry than humanity (it may be organic or perhaps it may be non-carbon based). Regardless, whatever the building blocks of their "super-nature" are, they remain corporeal and capable of existing in both the heavenly and earthly realms. The essential difference comprises the ability to exist in either realm and to move back and forth between the many layers of the "multiverse" like walking from one room to another. Having used that analogy, however, let me quickly clarify I believe this angelic capability isn't really about traveling from one dimension to the other; it is more about existing in both dimensions natively, due to the constituent parts that make them what they are. In other words, this does not mean angels can *travel* to higher, or extra-dimensions. It means they already take up space in those higher dimensions. Just as we do not travel from any one of the three dimensions we inhabit (height, depth, and width – usually possessing a bit too much of the latter), we remain a part of those dimensions concurrently. Likewise, angels exist within higher spatial dimensions even as they exist within lower dimensions at the same time; and they can be discernable by us in certain circumstances they control by their "native" functional capabilities. Obviously this concept constitutes a difficult notion to grasp, let alone work with it in a practical way. Thankfully, an author in the late 1800's provided us with an excellent thought exercise to help us appreciate how these ideas can be so.

The book *Flatland: A Romance of Many Dimensions* by A. Abbott (1884) explores the idea of a fictional two-dimensional universe called Flatland. The inhabitants of Flatland are known as Flatlanders and are described as simple shapes, such as squares, circles, and triangles. The story explores the question of how a being of three spatial dimensions would describe himself to, or interact with, a Flatlander and be properly understood by that two-dimensional denizen.

A Flatlander would have no concept of up or down. To a Flatlander, all that exists is forward, backward, left, and right. Because of this, height and depth are completely foreign concepts that Flatlanders would not be able to perceive or even envision in their minds. Even if a three-dimensional object were to breach their two-dimensional space, Flatlanders would only see it in their two-dimensional perspective.

If I were to drop a ball through Flatland, the Flatlanders witnessing this event would not see the ball as we do. They would see a point in space appear in front of them just as the ball breaches their space. Next, as the ball passes through, Flatlanders would see a circle (as a mere line) growing in size in front of them. The circle would reach its largest when the middle of the ball passes through. After that, as the ball is exiting, Flatlanders would see the circle shrink in size and eventually disappear completely. Technically, the Flatlanders did see the entire ball, only they saw it in numerous two-dimensional slices. Moreover, they can't meaningfully explain what they've just experienced. We can apply this to our three-dimensional experience as a thought exercise to help us imagine extra spatial dimensions (to be clear, we are only speaking of *spatial* dimensions here, not temporal; time as the fourth dimension is not a factor in this exercise).

What if a ball consisting of four spatial dimensions rather than three, called a hypersphere, were to breach our three dimensions of space?

For us, it would appear to be the same thing, but with an undetectable added dimension. We would first see a point in space appear. As the hypersphere moves through our space, the point would become a ball and continue growing in size. Once the middle of the hypersphere is through our space, we would see the ball start to shrink in size until it disappears completely. As a quick side-note, could it be significant that many reports of UFOs contain details such as these?

INTRUDERS FROM HIGHER DIMENSIONS

We can use this mental exercise to help us consider the extradimensionality of angels. An angel can breach our space and even look perfectly human, but that does not mean that's all there is to the angel. We would not be able to see what is hidden in the higher dimensions. Consider again the hypersphere. If the hypersphere were to stop midway through our dimensions of space, it would appear as a normal ball to us. If it were small enough, we would be able to touch it, pick it up, throw it around, and even fully enclose it in a container. However, there is one catch to this. If the ball was really a hypersphere, it should be heavier than we would predict based upon our preliminary observations.

Back to Flatland: imagine Flatland is vertical instead of horizontal. Imagine Flatland consists of left, right, up, and down, but not forward and backward. In short, imagine Flatland as a wall instead of a floor. This would mean Flatland would be subject to gravity. Even if Flatlanders were the same size as us by way of height and width, they would still weigh considerably less than we do because there would be *less* to their physical makeup. If a human were to step halfway through Flatland, he could look like a normal Flatlander to any other Flatlander. If Flatlanders were to try and weigh the human or pick him up, however, they would quickly realize something was different. They would discover he weighs a lot more than they do. The difference is the rest of his physical body is hidden in the third spatial dimension. Keep this in mind later when you read in Woodward's essay on the uncanny nature of "molecular deformation" which mimics one aspect of alchemical thought. Depending upon its molecular composition gold can weigh less than nothing! (Which might explain why the Ark of the Covenant was so easy to lift and carry.)

Looking at this from the other end, it would also mean the human subject would be incredibly stronger and more powerful than any Flatlander. Lifting a Flatlander would be no problem at all to a human. Even if the Flatlanders were to become frightened and attacked the human, the human would win the battle easily by sheer strength utilizing nothing more than the "extra" third dimension. The human could step into the third dimension and take the Flatlanders out, one by one, with ease. (A possible explanation for "alien abduction"?) The Flatlanders would have no defense. What the human did defied all the rules of their universe. Scaling this up to our three dimensions and considering extra spatial dimensions, this might help explain some of the amazing strength and abilities angels possess in the Bible:

> *That night the angel of the LORD went out to the Assyrian camp and killed 185,000 Assyrian soldiers. When the surviving Assyrians woke up the next morning, they found corpses everywhere. (2 Kings 19:35 NLT)*

EXTRADIMENSIONAL GENETICS

Laying aside the exact mechanics of how such a thing could actually happen (remember, this is just a thought exercise), imagine the human were to impregnate a Flatlander female. Most of the fertilization process happens on the Flatlander side of the encounter, so for it even to be possible, the three-dimensional human seed would have to somehow adapt to a two-dimensional world. In other words, if conception occurred, it would be within a two-dimensional environment. The genetics from the male would still be passed to the offspring, but the new body that is forming within the Flatlander would be two-dimensional. The Flatlander would be two-dimensional herself and would not be capable of housing, creating, or

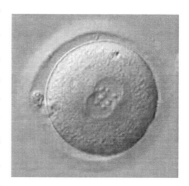

Figure 20 - A SINGLE CELL ZYGOTE, GAMETES JOIN

even contributing to something three-dimensional. Therefore, the three-dimensional reality coming from the seed would have to adapt to the new two-dimensional environment of the egg. We can scarcely imagine what this would do to the molecular structure of the zygote.

This raises the issue, however, that there are still three-dimensional traits being passed down to a two-dimensional child. Most of these, such as skin, hair, and eye color, wouldn't pose any problems. However, the genes that determine size certainly would. The genes determining three-dimensional size would be present in this new two-dimensional hybrid. Simply speaking, all of that mass and size has to go somewhere. With no other alternative, the size of the hybrid grows in the two dimensions it can actually inhabit, yet since there is so much more to its nature, the hybrid might grow to an incredible height and width. Once born, this two-dimensional Human-Flatlander hybrid would be seen (quite obviously) as a giant among all the Flatlanders.

This is all speculative of course, but if we employ that thought exercise considering three spatial dimensions, it might help explain why the Nephilim were "super-sized" humanoids. The extra-dimensional genetics of an angel would be passed to the hybrid child. The genetic encoding from "the father" that determines size would be present as well, but would have to conform to only three dimensions rather than four or more. Since there exists at least one less spatial dimension in which it can grow, the hybrid child would develop in the three dimensions it can inhabit in this abbreviated reality of three dimensions. If any veracity exists in the fourteenth-century citation below, our discourse could help explain this horrific text describing an abominable birthing process:

> [Women] with whom the angels had companied conceived, but they were unable to bring forth their children, and they died. And of the children who were in their wombs some died, and some came forth; having split open the bellies of their mothers they came forth by their navels. And when they were grown up and reached man's estate they became giants, whose height reached unto the clouds... (Kebra Nagast: Chapter 100)

Even if we discount the final statement as hyperbole (in other words, let's express our doubt that the giants really reached "unto the clouds"), – a human woman withstanding the birthing of a baby whose eventual height might be twice the size of a full-grown human – would require some odd concessions to its manner of delivery, to say the least.

MOTIVATIONS OF EXTRADIMENSIONAL ALIENS

This might also explain why so many alien abduction accounts include details about the entities being interested in our genetics. To be clear upfront, I am of the belief that alien beings are actually extradimensional by nature rather than extraterrestrial. I do believe they have the ability to engineer three-dimensional living bodies, however I believe the overall origin of these things are not from another planet, but rather from a higher dimension. Whether or not this "engineering" is through a sexual encounter or through technological means (think "test tube babies" as in Huxley's *Brave New World*), the outcome remains the same.

Therefore, it would seem that if copulation between an angel and a human be possible, gigantism would be the expected result. It would also seem angelic genes could carry traits not normally found in humans.

Figure 21 – BY LUCAS CRANACH THE ELDER (LANDESMUSEUM)

These traits could resolve into any number of superhuman abilities. Looking into tales of the ancient demigods could supply an idea of what those traits might be. We won't go down that path here, however.

Perhaps the so-called "alien entities" continue the attempt to create hybrids, combining their genetic material with humans. That possibility is more sinister than it is farfetched based upon the biblical texts we've considered in this book. Perhaps the ongoing process seeks to suppress problematic genes such as gigantism, while bringing more favorable genes to the surface (the "phenotype" to use the language of genetics[1]), such as those of telepathy, telekinesis, and perhaps even interdimensional travel (in light of our earlier discussion). This may all culminate in these extradimensional beings perfecting their malevolent plan (launched during the ancient, almost pre-historic past), unleashing it on our unsuspecting, skeptical, and mostly secular world.

SHAPESHIFTING

When looking at the Nephilim from an extradimensional perspective, perhaps predictably, the topic of shapeshifting arises. The most famous shapeshifter in our culture undoubtedly is the werewolf. This notion stretches back into the worship of wolves at Mount Lykaion in Greece[2] (from which we derive names like Lycos and a mental disorder known as Lycanthropy – thinking oneself a beast). Similar to werewolves, there are ancient legends of skinwalkers. In these instances, a person gains or is given the ability to change his or her human form to that of an animal. There are even stranger reports in modern times of individuals claiming to witness extraterrestrial beings shift from one shape to another, usually posing as a loved one or an animal. And then there is the legend of the Reptilians – extraterrestrials (or "inter-dimensionals") who look human until you look deeply into their eyes and you notice that their pupils are slitted like reptiles, instead of rounded as non-reptilian eyes are.

---

[1] "The observable physical or biochemical characteristics of an organism, as determined by both genetic makeup and environmental influences." – *FreeDictionary.com*.

[2] "Mount Lykaion was sacred to Zeus Lykaios, who was said to have been born and brought up on it, and was the home of Pelasgus and his son Lycaon, who were said to have founded the ritual of Zeus practiced on its summit. This seems to have involved a human sacrifice and a feast in which the man who received the portion of a human victim was changed to a wolf, as Lycaon had been after sacrificing a child." See Wikipedia: https://en.wikipedia.org/wiki/Mount_Lykaion.

This begs the question, how could something like this be physically possible? Most would say, "it can't" and the phenomena is passed off as illegitimate. On the surface, this would seem rational. After all, it seems physically impossible that something of a human size could shift into something as small as a bird, as some Navajo legends claim.

So where does the extra mass go? How can a skeletal structure bend and contort in the way needed for shapeshifting to happen? The same questions arise when considering how a being of human size can shift into something larger and more massive than itself. Where does the extra mass come from?

THE TWO TYPES

When it comes to shapeshifters, there are two types. The first is a being who has this ability natively ("baby you were born this way"). In ancient legends, this being was seen as a god or demon, but completely supernatural in origin. Today, such creatures are mostly regarded as aliens.

The second type is a human being who has been given the ability to shapeshift. This is usually due to a curse pronounced over an individual or perhaps an infection spread by a bite. The infection, such as with werewolves, comes from another shapeshifter while curses originate from humans believed to be in touch with the spirit world, such as shamans or mystics, and possess magical abilities to manipulate matter or cast a spell on their victim.

NATURAL SHAPESHIFTERS

The first type is a bit easier to understand than the second from a physics point of view. For this, as with most things of this nature, we can use the example of Flatland, a fictional two-dimensional universe, to help explain what might be happening. As I have stated in my books *Quantum Creation* and *Cherubim Chariots*, I believe there exists in creation a variety of extradimensional beings. From a Christian perspective, we might most recognize these beings as angels. As stated earlier, the example of Flatland is a great aid to help us understand higher dimensions and extradimensional entities.

Imagine Flatland as a vertical universe inhabiting your living room. This version of Flatland consists of only two spatial dimensions: up/down and right/left. It lacks forward and backward. The denizens of Flatland, known as Flatlanders, can only see and interact with things in those directions. This means, if you are standing in front of Flatland looking straight at it, the Flatlanders would have no way of knowing you were observing them. You would be in the third spatial dimension, and Flatlanders lack the ability to perceive or envision this extra dimension at all. To a Flatlander, you are an extradimensional being. And that would literally be true.

Now, imagine you were to wait until no Flatlanders were looking and you step halfway through Flatland. Once you were noticed, Flatlanders would assume you were two-dimensional like them. Even though you are a being composed of three-spatial dimensions, Flatlanders would lack the ability to detect the third dimension. Therefore, they would only observe the two-dimensional section of your body breaching Flatland.

Having access to the third spatial dimension would give you additional abilities Flatlanders lack. Imagine changing your position in Flatland. Instead of standing up vertically in the middle of Flatland, imagine lying on the ground so that you inhabited Flatland bisected at the waist. To the Flatlander, you would look extraordinary. From their vantage point, they would first see a normal enough Flatlander. However, once you began to move, they would stare in disbelief as you transformed from Flatlander into a small oval on the ground (which would appear to be nothing more than a two-dimensional section of your body). To a Flatlander, you would be a supernatural shapeshifter.

Every shape that you could shift into while in Flatland comprises a part of your physical makeup. The shapes are merely two-dimensional slices of your body. You could even appear as two objects to Flatlanders. If you stepped out of Flatland and left your arms in, Flatlanders would assume you have transformed into two floating circular shapes. The number of shapes you could shift into would only be limited to the number of positions you could muster and the number of two-dimensional slices by which you could be "divided" when intersecting Flatland. It truly would be a number far too large to estimate accurately.

Conceivably, the same thing could happen with real shapeshifters (if we grant them to be real). As demonstrated in *Quantum Creation* and *Cherubim Chariots*, some alien beings that human subjects purportedly witness are more likely to be extradimensional beings rather than extraterrestrial.[3] If an alien being is shapeshifting, it might be a clue that it is actually composed of four or more spatial dimensions. The extradimensional being would be operating in the fourth spatial dimension, "shape shifting" in our space, just as one would be in the third spatial dimension when (hypothetically) changing shape in Flatland. This may explain reports of alien beings changing into a familiar form of a loved one, a cherished pet, or a ferocious animal. It may seem outlandish (there's a pun in there somewhere), but the shape-shifter could be utilizing what amounts to three-dimensional slices of their own extradimensional physiology – a behavior we would regard to be totally supernatural, but which they would accomplish as simply as turning one's head.

CREATED SHAPESHIFTERS

This next example requires even more imagination to contemplate. Using the Flatland example, this conjecture would be comparable to giving a Flatlander the ability to utilize a third spatial dimension. I imagine it could be accomplished in two different ways.

The first isn't quite as mysterious as the second. Imagine again the vertical Flatland in your living room. Imagine you were to grab a Flatlander and turn him ninety degrees into the third spatial dimension. To the other Flatlanders, this Flatlander would appear to have shapeshifted into a vertical line. We could also alter the shape the Flatlander shifts into by bending him in different ways, utilizing the third spatial dimen-

---

[3] A quick side-note: I believe there are beings composed of three spatial dimensions that are truly alien; for example, certain beings could be more-or-less manufactured, on and off-world, by extradimensional entities, e.g., alien "grays" may be no more than biological robots.

sion. The main distinction here is the Flatlander would be more restricted on how many shapes into which he could shift because he (or she) lacks an entire spatial dimension.

While we are still conjecturing, we might even be able to teach the Flatlander how to bend into the third dimension on his (or her) own. Going even further, we might be able to teach the Flatlander how to enter completely into the third spatial dimension. The Flatlander would still consist of two spatial dimensions, meaning he would still have a two-dimensional perspective. As he moved through the third dimension, the Flatlander would still only see it in two-dimensional slices. In other words, he would see everything as a vertical line changing colors and shadow. However, to other Flatlanders, he would have the ability to disappear into a higher dimension and shapeshift. He would be looked at as, well… a god.

The second pathway amounts to something far more mysterious. Essentially, it involves the ability to turn a Flatlander into a full-fledged three-dimensional being. It would mean we somehow have the ability to grant this "chosen" Flatlander an extra spatial dimension so she would become a three-dimensional being with a three-dimensional perspective. Such a transformation would be quite miraculous and would seem as far over the top as one could go.

There is an ancient Navajo legend of persons being transformed into a skinwalker. This creature is said to able to transform into many different shapes. It would seem the person so affected started out as a mere human being, yet through some means (incidental to us here) became an extradimensional being. Conceivably, extradimensional technology might exist that could do this, but there might be a simpler explanation.

We could suppose any true extradimensional being has a multitude of three-dimensional shapes at their disposal; at least as many as we have two dimensions. Given this postulate, any three-dimensional shape, regardless how complex it might seem to us, would be easy work for a being encompassing four spatial dimensions. Again, it's equivalent to we three-dimensional beings considering shapes comprising but two dimensions. Any two-dimensional shape we select will still be far less complex than any three-dimensional shape that makes up our reality.

This might explain the skinwalker phenomena. Imagine an unfortunate Flatlander cursed by another not-so-considerate Flatlander – and we were witness to this tragedy from a third spatial dimension where we go unnoticed. Now imagine that we were motivated (perhaps out of compassion for his predicament), to dispose of the "cursed" Flatlander, then take his place. If we positioned ourselves exactly right (where we looked just like the Flatlander we had dispatched for parts unknown… but still in Flatland of course), the other Flatlanders would assume we (well one of us) was now the victim. Afterwards, we could keep shapeshifting – remorselessly terrorizing their community to no end (they deserved it after all, since they curse their citizens from time to time) – while every Flatlander informed of the background to this transformation (and now being haunted) were certain their cursed friend was responsible for this sudden series of frightening and unfortunate events.

Could this be why something similar to this hypothesized occurrence happened where "skinwalker phenomenon" are known to have taken place? Could this also help explain the "doppelganger phenomenon"? Perhaps. Then again, it might only be the "Bizarro World" effect from DC Comics (first appearing in April, 1960) in which the superheroes live on a cube called "htraE" – Earth spelled backwards)

CONCLUSION

The notion of extra spatial dimensions could be the key to understanding many supernatural phenomena. When we start thinking of spiritual beings as super-physical rather than wispy and ethereal, the strange and mysterious encounters that permeate some people's lives become a bit more plausible. In fact, from a physics standpoint, if extra dimensions do exist and entities dwell there, we should *expect* these types of things to take place.

As I have communicated in my books, interviews, and articles, sincere people can have vivid and inexplicable experiences that deserve a reasonable explanation. Indeed, plausible explanations are all the more "reasonable" when such experiences are examined in light of a biblical worldview. As it stands, however, the majority of Christians today, especially those who consider themselves intellectually superior to most everyone else, have little interest in contemplating these things. It is my sincere hope that all Christians, especially those who regard themselves committed to a high view of biblical inspiration, will not disregard the supernatural worldview of the Bible when they encounter the highly strange experiences that they, or someone they know and trust, may testify to in the days ahead. If so, together we can recover a good amount of territory stolen by the enemy through our failure to consult the Word of God and believe what it teaches us about living out a life of faith in God's amazing Cosmos.

Figure 22 - BIZARRO WORLD SUPERMAN ADVENTURE COMICS - 1960

THE DIVINE COUNCIL, LESSER ELOHIM, AND THE GOVERNMENT OF GOD

Gonzo Shimura and S. Douglas Woodward

AN INCONVENIENT TRUTH ABOUT HEAVEN

THE BIBLE IS A SUPERNATURAL BOOK. AS SUCH, MUCH WITHIN ITS PAGES COMPRISES THE UNEXPECTED. FOR CHRISTIANS SCHOOLED IN THE BIBLE, THERE CAN BE SURPRISING TRUTHS we totally miss because we have been literally indoctrinated with conventional beliefs that contradict what the Bible actually teaches. This is most inconvenient. It upsets the apple cart of what we believe. It can make us very uncomfortable.

When we open the door to new understandings like this, we can walk through it circumspectly (watching our step carefully as we are venturing on new ground), or we can slam the door shut. We can embrace a new way to comprehend the nature of God and His Cosmos, or we can leave well enough alone and turn our back to the highly strange – denying we ever glimpsed a different perspective to appreciate better what the Bible teaches. Perhaps there is no greater example of this unnerving experience than when we find out what the Word of God teaches about the governance of God, the elohim who reside with God, and their duties.

IS THERE ONE AND ONLY ONE *ELOHIM*?

When we hear the word *elohim*, we immediately assume that it refers to the LORD GOD, to Yahweh, or as He is sometimes called in an English speaking world, Jehovah. We don't think that *elohim* could in fact reference other "gods". And we certainly don't ascribe to polytheism, the view that there are many gods. That would be anathema. We have been taught that there is one God. If we are a modern-day Jew but otherwise orthodox in our belief, we leave the nature of God at that and venture no further. If we are orthodox Christians, however, we proclaim there is indeed only one God, one *elohim known in three persons* (and we capitalize the word, *Elohim* in reverence). Of course, this is the doctrine of the Trinity. While implicit in many scriptures in the New Testament, it isn't anywhere set forth precisely the way we wish it would have been to set the record straight.

Most of us are well aware that the early church grappled with the notion of "three-in-one" for its first 300 years. At Nicaea in 325 A.D., the matter was settled to the satisfaction of a vast majority of those 318 church patrons there (1800 in total were invited by Emperor Constantine).[1]

Nevertheless, in some regards, the Church has remained uneasy about the creed hammered out by the Church Fathers so long ago. Indeed, it has wrestled with conceptualizing the Trinity down to this day.

Figure 23 - FRESCO OF THE COUNCIL AT NICAEA IN THE SISTINE SALON

---

[1] The notion that the idea of the Trinity did not exist until Nicaea is patently false. "The council of Nicaea dealt primarily with the issue of the deity of Christ. Over a century earlier the term "Trinity" (Τριάς in Greek; *trinitas* in Latin) was used in the writings of Origen (185–254) and Tertullian (160–220), and a general notion of a 'divine three', in some sense, was expressed in the second century writings of Polycarp, Ignatius, and Justin Martyr. In Nicaea, questions regarding the Holy Spirit were left largely unaddressed until after the relationship between the Father and the Son was settled around the year 362. So the doctrine in a more full-fledged form was not formulated until the Council of Constantinople in 360 AD, and a final form formulated in 381 AD, primarily crafted by Gregory of Nyssa." See https://en.wikipedia.org/wiki/First_Council_of_Nicaea.

It is in this unsettled setting that we offer a further complication to the conception of God in Heaven and how He chooses to rule His creation, specifically overseeing humankind on this Earth. At the outset, the authors of this book affirm without equivocation, we believe in one God in three persons. We are Trinitarians through-and-through. And yet, we note the Bible expresses that there isn't just one elohim, there are many elohim. How can this be? It turns out just as we discussed the fact there are many angels that have had a profound impact upon the Cosmos and life upon this Earth, *there are many so-called elohim that rule with God in Heaven.* We can't be exactly sure whether they are angels or creatures surpassing an angelic nature. We do know this: ultimately, they are creatures with personality, free will, and authority. And what the Bible tells us about them does not fit the usual frame of reference most find so familiar. We will, however, cover in this chapter such beings exist alongside God in the "third heaven". The Bible calls them **elohim** – and they have jurisdiction over the Earth as God decreed.

While serving as the guidebook for our personal lives, a careful reading of the many narratives found in the Holy Scripture discloses the appearance of various other-worldly beings demonstrating cosmic dimensions beyond everyday experience. Many scenes in the spiritual world are often dismissed as allegorical or poetic, perhaps even irrelevant to our day-to-day Christian walk. However, when we encounter the truths found on the topic of *The Divine Council*, our understanding of Christian theology can get reordered. A passage that mattered little before because we deemed it obscure, suddenly becomes meaningful when the Divine Council is taken into consideration. And while we cannot address every nuance of the Divine Council in one chapter, we are able to highlight how it adds value in no small way to an essential realization of the Cosmos from a deeply biblical point of view.

MEETING THE DIVINE COUNCIL IN HEAVEN #3

Let's begin in home territory, starting with the commonly heard phrases "heavenly host" and "hosts of heaven" that refer to beings with whom God chooses to keep company. While familiar, do you know that these phrases may represent divine beings and not just stars or angels?

> *Thou, even thou, art LORD alone; thou hast made heaven, the heaven of heavens, **with all their host**, the earth, and all things that are therein, the seas, and all that is therein, and thou preservest them all; **and the host of heaven worshippeth thee**.* (Nehemiah 9:6, emphasis added)

The Hebrew word for *host* means "army... divisions (of an army); as a title of God: of Hosts (the heavenly armies), the Almighty, with a focus on great power to conquer or rule..."[2] The word for "heaven" has multiple levels of meaning. According to one Aramaic and Hebrew dictionary, it means *"region above the earth...place of the stars, sky, air...the invisible realm of God."*[3] Notice this definition covers three distinct places. First, it specifies the skies and atmosphere that envelope the globe on which we live. Secondly, it implies outer space where the stars reside. And lastly, it includes the invisible realm beyond, where God and countless angels reign. To enter this location of the Divine Council we must imagine ourselves transported to this *"third heaven"*.[4]

Consider then this scene from 1 Kings, chapter 22 verses 19-23.

> [19] And he said, "Hear thou therefore the word of the LORD: I saw the LORD sitting on his throne, <u>and all the host of heaven</u> standing by him on his right hand and on his left."
>
> [20] And the LORD said, "<u>Who</u> shall persuade Ahab, that he may go up and fall at Ramoth-gilead?" And one said on this manner, and another said on that manner.
>
> [21] <u>And there came forth a spirit, and stood before the LORD</u>, and said, "I will persuade him."
>
> [22] And the LORD said unto him, "Wherewith?" And he said, "<u>I will go forth, and I will be a lying spirit in the mouth of all his prophets.</u>" And he said, "Thou shalt persuade him, and prevail also: go forth, and do so."
>
> [23] Now therefore, behold, the LORD hath put a lying spirit in the mouth of all these thy prophets, and the LORD hath spoken evil concerning thee.

---

[2] GK H7372 | S H6635 צָבָא ṣāḇāʾ 487x "צ," Kohlenberger/Mounce Concise Hebrew-Aramaic Dictionary of the Old Testament, n.p.

[3] GK H9028 | S H8064 שָׁמַיִם šāmayim 421x "שׁ," Kohlenberger/Mounce Concise Hebrew-Aramaic Dictionary of the Old Testament, n.p.

[4] Paul traveled to the "third heaven" in 2 Corinthians 12:2.

Let's analyze this account verse-by-verse. First, we see the LORD sitting on His throne with the Heavenly Host surrounding Him on His right and left sides. This heavenly setting will become a standard motif when addressing the Divine Council. It also depicts *a third heaven* where God's throne resides (2 Corinthians 12:2). While we normally imagine spiritual realities dwelling in a misty, ethereal, non-physical existence, the scriptures say otherwise. The hosts of heaven "standing" at the right and left of God conveys physicality. Furthermore, the host themselves appear to have varying personalities as we uncover next.

In verse 20, a very peculiar exchange between the LORD and His host takes place. The LORD asks which member of the host would carry out His judgment on Israel's King Ahab which the LORD established one chapter earlier. To grasp why the LORD had decreed a severe punishment, we must first learn of the circumstances leading to His decree.

King Ahab attempted to negotiate an exchange to purchase the vineyard owned by Naboth the Jezreelite. But Naboth refused the offer, having previously received specific direction from God *not* to relinquish Naboth's rightful inheritance. Dejected, King Ahab went into what can only be termed as a "pity party" refusing even to eat. His wife Jezebel then took matters into her own hands, writing a letter in Ahab's name to the elders of the city where Naboth resided. The commands were extreme: bring a false charge against Naboth and have him executed. The elders did as Jezebel instructed through her deceptive letter. Therefore, Naboth would die from stoning under false pretenses. As a result, God decreed an equally severe judgment on Ahab. We read this account from the previous chapter in 2 Kings 21:

> *[17] And the word of the LORD came to Elijah the Tishbite, saying,*
>
> *[18] Arise, go down to meet Ahab king of Israel, which is in Samaria: behold, he is in the vineyard of Naboth, whither he is gone down to possess it.*
>
> *[19] And thou shalt speak unto him, saying, Thus saith the LORD, Hast thou killed, and also taken possession? And thou shalt speak unto him, saying, Thus saith the LORD, In the place where dogs licked the blood of Naboth shall dogs lick thy blood, even thine.* (2 Kings 21:17-19)

At the beginning of chapter 22, Jehoshaphat asked Ahab to go with him into battle at Ramoth-gilead to recover land belonging to Israel. Ahab's decision sealed his fate. Returning to where we left off, verse 20 of 2 Kings 22, we see God consulting with His Host on how best to carry out God's decree pertaining to Ahab. As mentioned a moment ago, individuals belonging to the Heavenly Host appear to possess distinct personalities. Yahweh asks "who" (implying personhood) would come forward. Verse 21 identifies one particular *spirit* who came forward and "stood before the LORD," speaking to the LORD.

Figure 24 – THE DEATH OF KING AHAB

Then verse 22 reveals the game plan offered by this particular spirit, who was to be a "...*lying spirit in the mouth of all his prophets,*" so that Ahab would agree to join with Jehoshaphat.

This might come as a surprise as it would seem unjust for God to make use of lies within His "judgment process". However, we know despite God's holy character, He can and does work His will through the shortcomings of sinful men and women. Ahab, through his wife Jezebel, broke no less than four of the ten commandments of God.[5] As the king of Israel, to say this was unacceptable behavior would be at best an understatement.

Furthermore, a New Testament passage reaffirms the same idea. We recall a memorable issue in Bible prophecy – *"the strong delusion"* which is, according to Paul, *sent by God* (2 Thessalonians 2:11). This *"strong delusion"* is directed at those who *"...might be damned who believed not the truth, but had pleasure in unrighteousness,"* (2 Thessalonians 2:12). In other words, God honors what such people desire when rejecting Him and choosing instead "pleasures of this world". By so doing, this does not make God the "father of lies," (a title specifically reserved for Satan – John 8:44). Rather, God only employs the fallen character of humanity to achieve His will. Similarly, the example recorded in 2 Kings 22 shows God weaving Ahab and Jezebel's deceptive tactics into His plan in order to assure what is good for Israel and the accomplishment of His promise. And finally, in verse 23, we see the LORD execute His judgment. The second portion of this passage from the KJV reads, *"...and the LORD hath spoken evil concerning thee."* However, we might be wise to prefer the translation of the English Standard Version (ESV), *"The LORD has declared disaster for you,"* or even the New Living Translation, *"For the LORD has pronounced your doom."*

When looking at the sequence of events, consider the role of the Divine Council in determining the deadly judgment of king Ahab:

1. Ahab and Jezebel directly opposed and disobeyed God's law.
2. God decreed judgment on Ahab, proclaiming he must die.
3. God turned to His heavenly council (Divine Council) to determine how exactly the judgment will be carried out.
4. A *spirit* volunteers with a plan acceptable to God.
5. The plan transpires on the Earth. Ahab believes the lie (as influenced by the "lying spirit") that leads to his death.

---

[5] "Thou shalt not murder...Thou shalt not steal...Thou shalt not bear false witness...Thou shalt not covet thy neighbor's house..." [Exodus. 20:13, 15-17]

You might be asking yourself, "If God is all-powerful, why would He need a council of divine beings or spirits to carry out His will?" And the simple answer is, "He doesn't." Rather, this is simply how God has chosen to conduct business. And it's shown right in the pages of the Holy Bible. To us, the most fascinating element of the narrative is how it discloses "the third heaven" where God resides on His throne, how He and the *lesser elohim* operate in this spiritual dimension, and how it later manifests in our physical dimension. Such passages cause us to wonder how often God has decreed events for our lives, and how He might have engaged these spiritual beings to carry out His decrees.

However, this represents only a minor example of the Divine Council. As we dig deeper into the topic, we will see how this truth runs throughout the supernatural biblical narrative from Genesis to Revelation. As a byproduct of this teaching, we will also see how our personal destiny through the grace of our Father, the work of the cross, and the resurrection of Jesus, has much to do with the Divine Council!

SONS OF GOD – MORE THAN ONE OF A KIND

The most startling aspect of the Divine Council paradigm is perhaps its explanatory power of the entire biblical narrative. Consider the following: God created Adam and Eve to live and tend to the Garden of Eden. They are blessed and told to increase and multiply, to fill the earth, and to subdue it (Genesis 1:28). Simply put, Adam and Eve are commanded to take Eden and replicate it across the Earth. God's idea of Eden was to create a habitat where humanity and God's Heavenly Host, both children of the Most High God, could remain members of one family, humanity ruling the Earth and the Heavenly Host, the Divine Counsel, overseeing the heavens.

When Adam and Eve partook from the tree of the knowledge of good and evil, however, humanity was separated from God. The whole point of the biblical narrative is to illustrate how God acts to bring humanity back into His family where we were meant to be. Despite what we might think, our designated place of residence is not to be solely on this Earth, but also in His presence in the spiritual dimension.[6] We will see in the pages ahead that the Divine Council has been corrupted too, although exactly when and how we do not know. We can only deduce that it was after the creation was completed and probably before humanity was. And the evil one, who deceived Adam and Even in the Garden, still roams freely awaiting a final, culminating manifestation like no other before the end of the present age.

> *Having predestinated us unto the <u>adoption of children by Jesus Christ to himself</u>, according to the good pleasure of his will.* (Ephesians 1:5)
>
> *To redeem them that were under the law, that we might <u>receive the adoption of sons</u>.* (Galatians 4:5)

If you've ever wondered why the New Testament uses adoptive language for those who accept Jesus as Messiah, the Divine Council narrative starts to supply more insight on this matter. The phrase or title "sons of God" is often attributed to "born again believers" in the New Testament (Romans 8:14, Galatians 3:26 etc.). The Bible tells us that believers are adopted into the family of God, and thus become "sons of God". This profound fact will become much more impressive as we delve further into the facts concerning the Divine Council in the pages ahead. But to do this, we must first turn our attention to the title "sons of God" in the Old Testament.

If Christians are the "sons of God", how is it possible that there were *redeemed* "sons of God" before the death and resurrection of Jesus? The answer is… there weren't any.[7] Rather, the title "sons of God" used many times in the Old Testament had a more specific application. There the phrase means something different. It conveys "*a direct creation of God, residing in the heavenly family and council of God*". In

---

[6] Note: Many atheists use the argument of evil to reject God. However, their attitude resembles closer to what the Bible says is true about humanity. Everyone knows that something is wrong with the world, and its because we all inherently know that there is a better place. The Bible tells us, that place is in heaven with God.

[7] This is not asserting that no one was saved in the Old Testament by the death of Jesus looking forward to his salvific work on the cross. The saints of the Old Testament would be redeemed ultimately by Christ's death. But in Old Testament usage, the "sons of God" do not refer to human beings.

this way, the first man, Adam (Note: specifically created – not evolved!) should be categorized like the angels as one of the "sons of God".

> ...*the son of Enos, the son of Seth, the son of Adam, <u>the son of God</u>.* (Luke 3:38)

Here we see Adam referred to as a "son of God". Does this make Adam equal to Jesus? The answer happens to be a bit complicated. On one hand, yes: Adam constitutes *the first man* while Jesus is referred to as *the second man* (1 Corinthians 15:47). In the context of Pauline teaching, it is due to the fact that Jesus reversed the sin that came into the world through Adam. While Adam was created in an innocent state (possessing a certain level of glory, likely even "gleaming" at his creation) he also lost his status. Therefore, Jesus comprises the redeemer of mankind, bringing us back into fellowship with the Father, redeeming us from the slavery into which we were sold (this is the picture of *redemption* from the Old Testament). In that sense, Paul calls Jesus the "second man" or "second Adam". Of course, we learn from Paul that the glory which God has for us, that awaits us at our resurrection, is a glory of the same sort that God's Son, Jesus Christ possesses. This goes far beyond innocence, and far beyond the glory which Adam possessed to a lesser extent when he was originally created. Indeed, this is a frequent theme of Paul. We see it in Romans and in 2 Thessalonians.

> *And if children, then heirs; heirs of God, and joint-heirs with Christ; if so be that we suffer with him, that we may be also glorified together.* (Romans 8:17)

> *Moreover whom he did predestinate, them he also called: and whom he called, them he also justified: and whom he justified, them he also glorified.* (Romans 8:30)

> *When he shall come to be glorified in his saints, and to be admired in all them that believe (because our testimony among you was believed) in that day.* (2 Thessalonians 1:10)

But the classic extended teaching on the glory that we are to receive we find within 1 Corinthians 15:35-55. Allow us to cite it fully here. Paul discusses the different "types of glory" and the distinguishing glory which we are to enjoy at the resurrection (aka rapture), when our bodies are transformed into their future, immortal, glorified state:

[35] But some man will say, "How are the dead raised up? and with what body do they come?"

[36] Thou fool, that which thou sowest is not quickened, except it die:

[37] And that which thou sowest, thou sowest not that body that shall be, but bare grain, it may chance of wheat, or of some other grain:

[38] But God giveth it a body as it hath pleased him, and to every seed his own body.

[39] All flesh is not the same flesh: but there is one kind of flesh of men, another flesh of beasts, another of fishes, and another of birds.

[40] There are also celestial bodies, and bodies terrestrial: but the glory of the celestial is one, and the glory of the terrestrial is another.

[41] There is one glory of the sun, and another glory of the moon, and another glory of the stars: for one star differeth from another star in glory.

[42] So also is the resurrection of the dead. It is sown in corruption; it is raised in incorruption:

[43] It is sown in dishonour; it is raised in glory: it is sown in weakness; it is raised in power:

[44] It is sown a natural body; it is raised a spiritual body. There is a natural body, and there is a spiritual body.

[45] And so it is written, The first man Adam was made a living soul; the last Adam was made a quickening spirit.

[46] Howbeit that was not first which is spiritual, but that which is natural; and afterward that which is spiritual.

[47] The first man is of the earth, earthy; the second man is the Lord from heaven.

[48] As is the earthy, such are they also that are earthy: and as is the heavenly, such are they also that are heavenly.

[49] And as we have borne the image of the earthy, we shall also bear the image of the heavenly.

[50] Now this I say, brethren, that flesh and blood cannot inherit the kingdom of God; neither doth corruption inherit incorruption.

> *⁵¹ Behold, I shew you a mystery; We shall not all sleep, but we shall all be changed,*
>
> *⁵² In a moment, in the twinkling of an eye, at the last trump: for the trumpet shall sound, and the dead shall be raised incorruptible, and we shall be changed.*
>
> *⁵³ For this corruptible must put on incorruption, and this mortal must put on immortality.*
>
> *⁵⁴ So when this corruptible shall have put on incorruption, and this mortal shall have put on immortality, then shall be brought to pass the saying that is written, Death is swallowed up in victory.*
>
> *⁵⁵ O death, where is thy sting? O grave, where is thy victory?*

However, there remains a difference from one "state" to another "state". Jesus exists as the one-and-only begotten Son (John 3:16), uniquely above all creation and the author of it (John 1:1-3). The Triune God is the Elohim of elohim. Jesus' "sonship" constitutes something quite distinct from ours. We are privileged to share in His glory as the greatest gift that the LORD could give any being. No doubt, this comprises a major part of the resentment and hatred that Satan holds toward humanity. Satan sought God's glory but he was not granted this glory nor did he have the means to acquire it. Instead, it was the destiny God established for humanity even before Adam's creation. Those redeemed by Christ will receive this glory at our resurrection.

The word translated "only begotten" in the Greek has some controversy surrounding its exact derivation and meaning. The two variants are "only offspring" or "only kind." We won't get into the technicalities here, but the point is this: Jesus was and remains categorically unique from the rest of the "sons of God" in that He *was* God. He was THE Elohim. The other elohim are not equal to Him. They are elohim but not co-equal nor co-eternal. *Jesus created them.* (Remember Woodward's treatment of this matter in the Introduction). Traditional theology which speaks to the unique identity of Jesus, remains intact even if the Divine Council is acknowledged. Its members are elohim, yet they remain "sons of God". So don't be confused. This is not an expression of polytheism.

With this foundation laid, we can now examine Old Testament instances which employ the phrase, "sons of God".

SONS OF GOD AND FUTILITY IN THE COSMOS

Earlier, we read in 1 Kings 21-22 about the death of King Ahab, but more precisely, the role God and His Heavenly Host played in Ahab's demise. Other parts of the Bible mention this "third heaven" setting too, including the presence of the "sons of God".

> *Now there was a day when the **sons of God** came to present themselves before the LORD, and Satan came also among them.* (Job 1:6)
>
> *Again there was a day when the **sons of God** came to present themselves before the LORD, and Satan came also among them to present himself before the LORD.* (Job 2:1)
>
> *When the morning stars sang together, and all the **sons of God** shouted for joy?* (Job 38:7)

The first two examples from the *Book of Job* show the "sons of God" presenting themselves before the LORD, much like we saw earlier the volunteer spirit come before the LORD. In Job 38:7, a clear distinction is made. The "sons of God shout for joy," celebrating God's fashioning of the earth (Job 38:1-7). The obvious indication: these beings existed prior to the creation of the Earth, and subsequently, humankind. Presumably, this opens up many questions about what the world was like before humankind appeared. Many theories exist ranging from "there was no world before humankind" to "there was another race that existed before Adam and we see evidence of this in archeology and geology". Specifically, those who identify themselves as the Biblical "gap theorists" argue that a pre-Adamic race existed; that is, a race of intelligent entities living on this Earth prior to Adam. According to Job 38:7, these sentiments seem to be confirmed. However, the exact state of their existence remains debatable. Until we are with Christ – until we are glorified – we will not know with any certainty the nature of these *sons of God* that "shouted for joy". Most theologians presume that these *sons* were angels. Might they have been the Divine Council? Conceivably, yes. Regardless, by definition these beings were outside the normal dimensions we experience if, in fact, the reality we call *space-time* was created *then*. However, if the world was a "re-creation" out of chaos caused by a previous rebellion, we might venture some other sort of being who did the loud exulting. Again, we don't know and the Bible doesn't supply any real hints.

The state of the universe prior to the Fall of humanity likely included incongruities of "what was in fact" with "what was originally intended". The Cosmos remains today in a state of disarray. This would explain why John's revelation speaks of a *"new heaven and new earth"* (Revelation 21:1), and a dramatic destruction of the physical world when he says the heavens receded *"like a scroll"* (Revelation 6:14, Isaiah 34:4 – although the scope of this event is not made clear), as this will occur when the effects of the curse on physical creation are lifted. In fact, the prophetic transformation of the church as the "sons of God" is connected to the Cosmos' own transformation when in Romans 8:19 it states that the creation *"waits with eager longing for the revealing of the sons of God."* We might speculate, and we think rightly, that the "sons of God" in Job 38:7 had some sort of physical state of existence. Regardless, they were certainly spiritual in nature above all else (see Josh's earlier explanation of the interconnection of angelic "spirits and their physics"). If the "sons of God" were present at the time of the creation of the Earth and even at the time when Yahweh created Adam and Eve, obviously this was prior to the corruption that came to infect the physical creation when it was subjected to "futility": [8]

> [20] *For the creature was made subject to vanity, not willingly, but by reason of him who hath subjected the same in hope,*
>
> [21] *Because the creature itself also shall be delivered from the bondage of corruption into the glorious liberty of the children of God.* (Romans 8:20-21)

---

[8] The timing of the subjection of the Creation to futility is usually thought to be connected to the fall of humanity with the sin of Adam and Eve and their expulsion from the Garden of Eden. However, it is conceivable it occurred before this rebellion. That *futility* ("pointlessness – without use") might be an equivalent statement regarding the creation being or perhaps more accurately stated as *becoming* "formless – and made void" as stated in Genesis 1:2. If so, it raises some interesting questions about whether the re-creation affirmed by those who subscribe to the "Gap Theory" contends that Creation's restoration was total – that every aspect of creation was returned to a pristine state OR only *selected portions as touched* by God in Genesis 1 were made over and made "good". This discussion will be considered more thoroughly in Volume 2 of *Revising Reality*.

After the fall, the next mention of the "sons of God" comes in Genesis 6. But in this account these sons now exist in rebellion, and appear to breach the boundaries established as the result of the fall.

> *¹ But there were false prophets also among the people, even as there shall be false teachers among you, who privily shall bring in damnable heresies, even denying the LORD that bought them, and bring upon themselves swift destruction...*
>
> *⁴ For if God spared not <u>the angels that sinned</u>, but cast them down to hell, and delivered them into chains of darkness, to be reserved unto judgment;*
>
> *⁵ And <u>spared not the old world</u>, but saved Noah the eighth person, a preacher of righteousness, bringing in the flood upon the world of the ungodly...* (2 Peter 2:1, 4-5)
>
> *And the <u>angels which kept not their first estate, but left their own habitation</u>, he hath reserved in everlasting chains under darkness unto the judgment of the great day.* (Jude 1:6)

Both Jude and Peter affirm the notion that there were angelic beings that did not keep their own habitation. In Peter's account, these angels that sinned were cast down to hell or Tartarus. In fact, 2 Peter 2:1

Figure 25 - SONS OF GOD – DAUGHTERS OF MEN
Daniel Chester French 1923, Corcoran Gallery of Art

explains how false prophets and false teachers would bring about destruction and judgment, just as it did for the "angels that sinned". After verse 4, Peter describes the days of Noah and the flood upon the world of the ungodly. It should not be overlooked that Peter writes of the *angels that sinned* immediately after he mentions *sparing not the old world in the days of Noah*. This links to the Genesis 6 account of those days of Noah *prior to the flood*:

> *And it came to pass, when men began to multiply on the face of the earth, and daughters were born unto them, That <u>the sons of God saw the daughters of men</u> that they were fair; and they took them wives of all which they chose. And the LORD said, "My spirit shall not always strive with man, for that he also is flesh: yet his days shall be an hundred and twenty years. There were giants in the earth in those days; and also after that, <u>when the sons of God came in unto the daughters of men, and they bare children to them, the same became mighty men which were of old, men of renown.</u>"* (Genesis 6:1-4)

Here we see the "sons of God" lusting after the daughters of men, even bearing children to them. These "sons of God" are made even more distinct from the daughters of men when we see that the Hebrew word for "men" in "daughters of men" is none other than "Adam," underscoring that "sons of God" and "daughters of Adam" are intentionally distinguished. As we have stated early on, no surprise then the mention of *"the angels that sinned and that kept not their own habitation"* by both Jude and Peter, explicitly references Genesis 6.

The Second-temple-period text found in the first 37 chapters of the *Book of Enoch*, supplies details the Genesis 6 account omits. Here, the "sons of God" are called *watchers*. This label is used in Daniel 4:17.

> *"This matter is by the decree of the watchers, and the demand by the word of the holy ones..."*

These watchers in the *Book of Enoch* are mentioned by name and by number; 200 that descended on Mount Hermon in Northern Israel (today's Golan Heights). The sins of these watchers are described in more detail, adding commentary on Genesis 6:1-4. The most repeated sin committed by the Watchers is that of mating with human women.

> At that moment the Watchers were calling me. And they said to me, "Enoch, scribe of righteousness, <u>go and make known to the Watchers of heaven who have abandoned the high heaven, the kodesh eternal place, and have defiled themselves with women, as theirs deeds move the children of the world, and have taken unto themselves wives</u>:
>
> They have defiled themselves with great defilement upon the earth; neither will there be peace unto them nor the forgiveness of sin. (1 Enoch 12:4-5)

The remarkable insight from Enoch reinforces the Divine Council paradigm. There are heavenly hosts, the "sons of God", some of which are known as the "Watchers", who may or may not be associated with God's Divine Council. As we will see, every member of the Council apparently has been corrupted. This raises a number of other questions about which we are left to wonder.

While we won't go into the details of the Nephilim here (having done so earlier), it ought to be noted another reason why this act was so abominable to God included the fact *it crossed dimensional barriers*. More than the mixing of species, when the "sons of God" mated with the "daughters of men", the offspring became hyper-dimensional entities that literally had one foot in each of the two realms.[9] It might explain why the word *nephilim* can be derived from "earth born," as in the Greek *Septuagint*.[10]

This race of beings possessed *supernatural* capabilities due to its extra-dimensional abilities. We are not talking about extraterrestrials mixing their genetics with humanity, although this is easily confused.[11] Rather, we are asserting that beings from the "other side" "interface" with humanity, begetting demi-god offspring as described in ancient mythology,

---

[9] A number of authors and experts who study this topic believe that when humanity sinned, our ability to interact with both realms (or multiple dimensions beyond those of space-time) was made dormant. The power became, as Watchman Nee claimed, the Latent Power of the Soul. Only a small percentage of humanity seem to have the ability to detect if not interact with this alternative realm. If the Nephilim were able to do this (and it seems likely they were), this would be an example of attempting to override God's wish to retard our ability to be "extradimensional", possibly out of hope that evil would be reduced.

[10] See https://www.khouse.org/articles/1996/43/

[11] Note: The *alien abduction phenomenon* as well as the alleged "hybrid breeding program" of aliens and humans might be amplified by the Divine Council paradigm.

or as the Bible states, *"mighty men who were of old, men of renown."* It is conceivable these beings *possessed* and *used* clairvoyance and psychokinesis – causing humanity in ancient times to equate them with gods.

NO LONGER SO KODESH – NO KIDDING

One of the major differences between spiritual beings called "sons of God" or "Watchers" and humans, is that the former were at least originally (and perhaps remain) immortal. There is something to the nature of the unseen dimension that exists apart from the systems of decay and death. And since death entered the world from sin (Romans 5:12), it was Jesus Christ whose death reversed this reality. Men and women experience death as part of our natural existence. But it wasn't so for these "sons of God" – at least not until they sinned. In Hebrew the word transliterated (pronounced) *kodesh*, means "holy or set apart, free from sin, sanctified". This word applies to the "sons of God" before they copulated with women (women who were *existing in a sinful state* as a result of humanity's earlier sin). The sons of God who descended on Mount Herman and then mated with the "daughters of men" ceased being *kodesh*.

Consider the following passage from 1 Enoch 15:

> *15:3 "For what reason have you abandoned the high, kodesh, and eternal heaven; and slept with women and defiled yourselves with the daughters of the people, taking wives, acting like the children of the earth, and begetting giant sons?"*
>
> *15:4 "Surely you used to be kodesh, spiritual, the living ones, [possessing] eternal life; but now you have defiled yourselves with women, and with the blood of the flesh begotten children, you have lusted with the blood of the people, like them producing blood and flesh, which die and perish."*
>
> *15:5 "On that account, I have given you wives in order that seeds might be sown upon them and children born by them, so that the deeds that are done upon the earth will not be withheld from you."*
>
> *15:6 "Indeed you, formerly you were spiritual, having eternal life; and immortal in all the generations of the world."*
>
> *15:7 "That is why formerly I did not make wives for you, for the dwelling of the spiritual beings of heaven is heaven.'"*

15:8 "But now the giants who are born from the union of the spirits and the flesh shall be called evil spirits upon the earth, because their dwelling shall be upon the earth and inside the earth."

15:9 "Evil spirits have come out of their bodies. Because from the day that they were created from the kodesh ones they became the Watchers; their first origin is the spiritual foundation. They will become evil upon the earth and shall be called evil spirits."

15:10 "The dwelling of the spiritual beings of heaven is heaven; but the dwelling of the spirits of the earth, which are born upon the earth, is in the earth."

15:11 "The spirits of the giants oppress each other; they will corrupt, fall, be excited, and fall upon the earth, and cause sorrow. They eat no food, nor become thirsty, nor find obstacles."

15:12 "And these spirits shall rise up against the children of the people and against the women, because they have proceeded forth from them."

There are numerous themes here that echo the New Testament in this excerpt from *Enoch*, and one interesting declaration emerges regarding the origin of demons (see footnote).[12] And notice the punishment of these fallen ones: the loss of life *in the heavenlies* for the "Watchers" resulting from their rebellion. Another set of divine beings receive the same verdict which we will investigate in a moment with Psalm 82.

After the flood, Noah's three sons began to repopulate the Earth. Genesis 10 outlines what is called the "Table of Nations". There 70 nations are listed as the resulting genealogies from Shem, Ham and Japheth. It is important to point out that the nation of Israel is *not* mentioned as one among the seventy. This will become an important point when we learn more about the *governance of God*. In Genesis 11, we read about the construction of the Tower of Babel, and the resulting dispersion of the human race as well as its separation into many different language groups (Genesis 11:9). Here the "sons of God" once again enter into the account.

---

[12] This passage is often used to explain demons in the New Testament. The theory goes that when the Nephilim were wiped out in the flood of Noah, their spirits were "loosed", becoming the demons seeking physical refuge in physical bodies – humans or other creatures. While demons were not mentioned often in the Old Testament, the presence of Jesus appears to "flush them out". They appear frequently in His ministry.

> *When the Most High gave to the nations their inheritance, when he divided mankind, he fixed the borders of the peoples <u>according to the number of the sons of God</u>.* (Deuteronomy 32:8 ESV)
>
> *When the Most High divided to the nations their inheritance, when he separated the sons of Adam, he set the bounds of the people <u>according to the number of the children of Israel</u>.* (Deuteronomy 32:8 KJV)
>
> *When the Most High gave the nations their inheritance, when he divided up humankind, he set the boundaries of the peoples, <u>according to the number of the heavenly assembly</u>.* (Deuteronomy 32:8 NET)

You should see why we cited three different translations of this text. The ESV clearly says that God divided mankind and fixed the borders of the people according to the number of the "sons of God". This aligns well with the narrative we recited from Genesis 11 in which God "confused" the people's tongues at the Tower of Babel. The KJV however, states the boundaries were set according to the "sons of Israel". This constitutes an anachronism since at the time when the Most High divided the nations (at the Tower of Babel incident) *the nation of Israel didn't exist!* (It would be hundreds of years later before it did). The NET (New English Translation) renders this passage much more in line with the ESV – "the heavenly assembly" – reflecting manuscripts retrieved from the Dead Sea Scrolls according to Dr. Michael Heiser. Dr. Heiser adds:

> Controversy over the text of this verse concerns the last phrase, "according to the number of the sons of Israel," which reflects the reading of the Masoretic Text of the Hebrew Bible (hereafter, MT) ... Literary and conceptual parallels discovered in the literature of Ugarit, however, have provided a more coherent explanation for the number 70 in Deuteronomy 32:8 - and have furnished powerful ammunition to textual scholars who argued against the "sons of Israel" reading in MT... Hence the point of Deuteronomy 32:8-9 is not merely that God created seventy territorial units after Babel, but that each of these units was given as an inheritance. The question is, to whom were the nations given? This is left unstated in Deut. 32:8a, but 32:8b, the focus of our controversy, provides the answer. The parallel only makes sense if the original reading of 8b included a reference **to other divine beings to whom the other nations could be given: the "sons of God."**[13] [Emphasis added]

---

[13] Heiser, Michael S., 'Deuteronomy 32:8 and the Sons of God' http://www.thedivinecouncil.com/DT32BibSac.pdf

As Dr. Heiser indicates, the more accurate reading in Deuteronomy 32:8 is "according to the sons of God" – upholding our grand narrative of the "sons of God" and the Divine Council. So when the languages were divided at the Tower of Babel, God didn't abandon humanity; rather He placed particular members of His Divine Council, other "sons of God" as *watchers* over mankind – seventy to be specific. This perspective fuels the possible explanation for the foundation of all religions around the world. That is to say, many share concrete principles, yet *their gods always differ*. What we learn in Psalm 82 is what became of these other gods – these lesser elohim, these "sons of God" who inherited the 70 nations as their spiritual rulers. But they dropped the ball.

> *¹ God standeth in the congregation of the mighty; he **judgeth among the gods.***
>
> *² How long will ye judge unjustly, and accept the persons of the wicked? Selah.*
>
> *³ Defend the poor and fatherless: do justice to the afflicted and needy.*
>
> *⁴ Deliver the poor and needy: rid them out of the hand of the wicked.*
>
> *⁵ They know not, neither will they understand; they walk on in darkness: all the foundations of the earth are out of course.*
>
> *⁶ I have said, **Ye are gods**; and all of you are children of the Most High.*
>
> *⁷ **But ye shall die like men, and fall like one of the princes**.*
>
> *⁸ Arise, O God, judge the earth: for thou shalt inherit all nations.*
> (Psalm 82:1-8, emphasis mine)

Here we see God judging the other (lower-case-g-plural) gods. They are called "children of the Most High," indicating the identity of these "gods" as synonymous with the "sons of God." God declares the assignments they failed to fulfill. He explains how these elohim judged unjustly, did not defend the poor, the fatherless and the needy. He exposes their ways of darkness and decrees judgment upon them. *"I have said, **Ye are gods**; and all of you are children of the Most High. But ye shall die like men, and fall like one of the princes."* The punishment decreed by God is consistent with the judgment we noted previously from the

Book of Enoch. These "gods" would lose their heavenly home, their status. They will "die like men" as will *the Watchers.*[14]

These 70 "sons of God" accepted worship from the people, rebelling against God's ordered decree. Could it be that some of them committed the same sin as the Watchers, recorded in Genesis 6 and the *Book of Enoch*?

"This number 70" may be tied to why Jesus sent out 70 disciples during His ministry to preach the good news throughout the land. We believe it symbolizes our destiny. We – as redeemed, reborn, and adopted sons of God – will replace the 70 rebellious Divine Council members, whose fate is everlasting darkness.

Despite these 70 rebellious "sons of God" that Yahweh declares will "die like men," He chose to work His messianic promise through His allotted nation – the nation he kept for Himself – Israel. Ultimately, many generations later, individuals from the disinherited nations would be brought back into God's family through the work of Jesus.

This cosmic drama underlies much of the Old Testament story… this battle for the souls of men between Yahweh and the rebellious gods. The outcome of this war presents two destinies for individual human beings; a place with God serving in a new Divine Council, or in the Lake of Fire with Satan and his angels. Keeping the "Divine Council paradigm" in mind, the implications of Deuteronomy 4:19-20 are made clearer:

> *And lest thou lift up thine eyes unto heaven, and when thou seest the sun, and the moon, and the stars, even all the host of heaven, shouldest be driven to worship them, and serve them, which the LORD thy God hath divided unto all nations under the whole heaven. But the LORD hath taken you, and brought you forth out of the iron furnace, even out of Egypt, to be unto him a people of inheritance, as ye are this day.* (Deuteronomy 4:19-20)

---

[14] Are all sons of God forever immortal? It would appear that "dying like men" infers that their end will be futility. They would be reduced to nothing in the metaphoric way that humans lose their status and "place in the world" upon death. They will be like princes falling from their throne. They will be ashamed and regarded with great disdain. However, Dr. Michael Heiser suggests the "council members" became mortal.

The essence of the idea repeats in verse 9 of Deuteronomy 32, right after the controversial passage we mentioned earlier with the translations between the "sons of Israel" and the "sons of God".

> *For the LORD'S portion is his people; Jacob is the lot of his inheritance.* (Deuteronomy 32:9)

Tradition teaches that Moses wrote Deuteronomy. If true, it would make sense that he would possess a panoramic picture of the cosmic battle between Yahweh and the lesser elohim. His firsthand experience with the gods of Egypt and their defeat at his hands empowered by the LORD (the great "I AM" in Exodus 3:14) would emblazon the impact of this spiritual conflict beyond what any other human being would experience (except possibly Elijah during his battle with the priests of Baal at Mount Carmel). There are nine generations between Adam and Noah. After the flood, there are ten generations from Noah to Abraham. From there, we have six generations before Moses.[15] The reference over and over again regarding God bringing Israel out of Egypt gains added significance due to the "backstory" of the Divine Council.

THE DIVINE COUNCIL AND YOU

So why does the Divine Council matter? Does it apply to members of the church of Jesus Christ? If you are a born again believer in the resurrected Christ Jesus, the answer is a resounding "YES!"

In his book cited in this chapter, *The Unseen Realm: Recovering the Supernatural Worldview of the Bible*, ancient Near East linguist and author Dr. Michael S. Heiser explains that the concept of the Divine Council runs like a red thread through the Old Testament:

> In the distant past, God disinherited the nations of the earth as His co-ruling family, the original Edenic design, choosing instead to create a new family from Abraham (Deuteronomy 32:8-9). The disinherited nations were put under the authority of lesser <u>elohim, divine sons of God</u>. When they became corrupt, they were sentenced to mortality (Psalm

---

[15] Map of Bible genealogy: http://www.actscambridge.org/assets/images/bible_genealogy_a1_chart.pdf

82:6-8). The Old Testament is basically a record of the long war between Yahweh and the gods, and between Yahweh's children and the nations, to re-establish the original Edenic design.[16]

Dr. Heiser also indicates how the Old Testament premise of the Divine Council paradigm intersects prominently with the life and ministry of Jesus Christ as recorded in the Gospels. In fact, the Divine Council has much to do with our blessed hope as believers with the return of Christ.

> The victory at Armageddon of the returning incarnate Yahweh over the Beast (antichrist) who directed the nations against Yahweh's holy city is the event that topples the <u>elohim</u> from their thrones. It is the day of Yahweh, the time when all that is wicked is judged and when those who believe and overcome replace the disloyal sons of God. The kingdom is ready for full, earthly realization under a reconstituted Divine Council whose members include glorified believers. The full mass of believing humanity will experience a new Edenic world in a resurrected, celestial state.

For a comprehensive and unquestionably scholarly treatment of the Divine Council, we strongly recommend studying the entirety of Dr. Heiser's work. It is a remarkable book.

Once we have grasped its implications, the Divine Council paradigm is a powerful lens through which we can view the scriptures. It provides an overlooked but potent means to understand the mind of the ancient Israelite, whose context is far different than ours and who thus understood the Word of God in a dramatically different way. What is the major thematic distinction? *Their supernatural worldview.* The Cosmos had been corrupted by evil – the rebellious elohim were led by Satan himself.

> *Again, the devil taketh him up into an exceeding high mountain, and sheweth him all the kingdoms of the world, and the glory of them; And saith unto him, All these things will I give thee, if thou wilt fall down and worship me.* (Matthew 4:8-9)

While much theology has sprung from this passage concerning the devil and his dominion over all earthly kingdoms, the Divine Council is not often brought into the cosmic narrative. The temptation of Jesus

---

[16] Heiser, Michael S., (2015) 'Chp. 42: Describing the Indescribable,' *'The Unseen Realm'* (pp. 376) Lexham Press, Bellingham, WA

in the dessert happened at the beginning of Jesus' ministry. And as the devil proclaims (and Jesus does not dispute it), at that moment he controls all the kingdoms of the earth. So the implication is that all 70 rebellious Divine Council members had rejected Yahweh and His decree to watch over the nations. They had pledged allegiance to one ruler, Satan. Since Satan offers Jesus the glory of the kingdoms of the Earth if He would only bow down to the Devil, it's not a stretch to suppose that the Devil's offer had the backing of the other members of the Divine Council, the original 70 Yahweh assigned 2,000 years earlier during the "Babel Event" and as later recorded in Deuteronomy 32:8.

So, our rightful place – our inheritance as redeemed humans – is to hold a seat in the newly reconstituted Divine Council, an amazing opportunity which lies before us. Along with new eternal bodies as Paul outlines in 1 Corinthians 15, we will be promoted into the council of God:

> *After this I looked, and behold, a great multitude that no one could number, from every nation, from all tribes and peoples and languages, standing before the throne and before the Lamb, clothed in white robes, with palm branches in their hands, and crying out with a loud voice, "Salvation belongs to our God who sits on the throne, and to the Lamb!"* (Revelation 6:9-10)

What is outlined later in Revelation 22 as the final restoration of the Edenic state of the world, signifies the completion of God's cosmic narrative for this eon, this particular given age. Having access to the river of the water of life and the tree of life, Yahweh and humankind are restored in the intended relationship God designed for us so long ago – before the world was.

When the Divine Council meets with its new members in place (millions and millions of them), we will be glorified, our adoption will be realized, and we will be co-seated with Christ in Jerusalem, His millennial capital. We will fully enjoy our inheritance in Christ and He will be overjoyed with His inheritance in us. Let us therefore…

> *(Look) unto Jesus the author and finisher of our faith; who for the joy that was set before him endured the cross, despising the shame, and is set down at the right hand of the throne of God.* (Hebrews 12:2)

THE ELECTRIC UNIVERSE
Anthony Patch and S. Douglas Woodward

*For my thoughts are not your thoughts,
neither are your ways my ways, saith the LORD.*
Isaiah 55:8

OCCULTED PHYSICS

TODAY, AS THROUGHOUT OUR KNOWN RECORDED HISTORY, KNOWLEDGE AND TRUTHS HAVE BEEN OCCULTED FROM MANKIND. HOWEVER, NEW LIGHT IS NOW BEING SHED UPON THE MYSTERIES kept hidden. Our focus here zeroes in on the topic of physics, with its own arcane language and discoveries held closely within its priesthood of practitioners. High on our agenda: to reveal for your critical evaluation and spiritual edification, the model – the mental framework – by which we've grown comfortable with the *presumption*, for that is what it is, that *we understand our universe.* Indeed, humanity possesses enormous overconfidence; and not just the common man, but the scientists too, many who have become "rock stars" in our day, all professing that we now know the nature of the universe.

And yet, there is an undercurrent of other great minds that challenges this confident orthodox outlook. And the key to uncovering a new way, an unconventional way to see our Cosmos, stands CERN, the *Large Hadron Collider – the* most sophisticated machine every built by humankind. Learning its secrets opens the door to radically new ways to think about reality. But its mysteries also remind us that not all ventured in the name of science constitutes what's good for humankind.

Since the Garden of Eden, humanity has struggled to be freed from the lies and deception promoted by Satan. We unashamedly acknowledge this truth. Our overall objective in this chapter is to loosen the chains through which the Deceiver has held us captive. It is not really meant to be a religious nor philosophical discourse; rather, we hope to provide an explanation of what lies at the core of our existence; that is, the Cosmos as designed and crafted by the hands of Jehovah God, our Creator, who we know through the person of Jesus Christ, the Logos by whom and through whom all things were made that were made. (John 1:3)

Within his physical realm we can discover the mechanisms that guide how things work, from the smallest particle to the mightiest of galaxies. These mechanisms, these rules of the Cosmos, are uncovered and codified by the science of *physics.* Despite the mountain of learning we've already stored in our "data bank", many rightly persist in an ongoing exploration into the unknown for good reasons. We discover that there are often limits to the established rules of the road previously worked out by our predecessors; and thus, we frequently have to revisit the rulebook we formerly held to be sacrosanct. We have to update or even throw away some of the old cherished truisms. Einstein did this with Newton. Others had done this with Einstein. And our contemporaries who may cast aspersions at these great minds that have gone before us because they failed to "get it right" are filled with foolish hubris, for the work will never be finished – at least not in this age dominated by the frailty of humankind generally uninformed by the knowledge of God. Not surprisingly, the limits of our knowledge generally lie in those matters in the realms of the smallest and the largest – where our senses, even those enhanced by technology, stumble. Still, we dig deeper and we reach out farther in hopes we will break the code of the Cosmos.

At the outset, allow us to say that it must be kept in mind one cannot separate the physical from the spiritual. Each plays an equal role in both our existence and our relationship with our Creator. Indeed, we do well to keep in mind that the division of reality into two halves is itself a mistaken concept. Separating the Cosmos into nature and supernature amounts to an admission that we still have much work to do, because ultimately the two are a unified whole. To say the least, the rules that unify these distinct realms are yet to be understood. From one side of the divide, science quests to discover the bridge between the two realms, while spiritual science – theology, once known as the queen of all sciences – too often feels threatened by this examination, seeing the secular quest into the realm of the spirit a means by which science will invalidate theology. Rightly construed, however, theology should not be threatened by any genuine search for knowledge motivated purely by inquiry. The knowledge of the universe – how it really works – teaches us more about the mind of its creator; that His ways remain above our ways. (Isaiah 55:8) To understand the Cosmos, if we would understand it truly, would be to understand a vast amount about God Himself. The creation teaches us much about the creator, for God meant that it would disclose His nature to every man and woman, so in the day of judgment they would have no basis to deny they knew about Him *"For the invisible things of him from the creation of the world are clearly seen, being understood by the things that are made, even his eternal power and Godhead; so that they are without excuse."* (Romans 1:20) Therefore, believers in spiritual truths need only be suspicious of those whose search sets out on a misguided campaign to prove falsehood built on flawed premises; assumptions about reality constructed to protect cherished, but ultimately malevolent beliefs residing in that darkness, untruth, and deception conceived and sold to us by Satan at humanity's beginnings thousands of years ago.

This exploration into our mental model of the universe centers on the applied physics at the heart of the Large Hadron Collider (LHC) in Geneva, Switzerland, specifically designed, built, and managed by the organization we know as CERN. However, a word of warning as we begin: spiritual forces are at work. These forces transcend the human agents there, many unwittingly engaged in a program not of discovery, but of destruction. The science at CERN combines that which is known by the physicist, but also what is ventured by the alchemist. To these practitioners, some of them most unholy, lies the challenge to open a vault into the pit of hell. Exactly what their ultimate motive consists of cannot be concluded cleanly nor proved to an impartial jury of their peers. Yet, for those who have the eyes of spiritual discernment, their wicked goal is not so occulted. We can readily discern they seek to accomplish what seems to the sane mind an impossible task: *to kill God.*

Figure 26 - SHIVA, THE GOD OF DESTRUCTION AT CERN

Such a goal could only come from the mind of the greatest criminal in the Cosmos – Satan. How his henchmen plan to achieve the unachievable in his name lies within what is written in the pages ahead.

EINSTEIN & TESLA: DISTINCT VIEWS OF THE COSMOS

Albert Einstein... almost everyone has heard of this man, acknowledged for his brilliance within the world of physics. He is often credited as the greatest physicist of all time. His name signifies genius. But Nicola Tesla? Far fewer know of this man. Fewer still, his prowess in physics. And yet, how many know that Einstein acknowledged Tesla as the *real* genius? Generally speaking, they were contemporaries and both public figures, but Einstein's star power far outstripped Tesla's. Today's "fringe science" believes in "zero-point energy", extols viable flying saucers which defy gravity, and champions traversing time through inter-dimensional portals, "navigable" by humanity. To these open-minded souls, Tesla was the prescient mad scientist discovering these things one hundred years ago.

Using these two brilliant theorists and practitioners of science as the pillars of two significantly divergent models, we affix two distinctive labels of *gravity and electricity.* To Einstein we assign the phenomenon we call gravity; to Tesla, the phenomenon we call electricity. Science knows that there is a relationship between them, a relationship orthodoxy labels *electro-magnetism*. Here we will cast doubt on this orthodoxy. The relationship between the two is not exactly what we have been taught. In fact, the truth will astonish you.

For starters, to the finely schooled majority, our world and the universe are viewed through the lens of gravity – we see it as a principle governing property of all that is. Einstein based his grand theory of relativity upon it. Notions we accept today as enlightened science, if the truth be told (and we intend to do that here) untethers all these truths – such things as relationship of mass and energy, the bending of space-time, the fixed speed of light, and the light-speed barrier which cannot be exceeded. We won't have time nor space (no pun intended) to loosen all these "facts" from their moorings. But we will discuss why they are all open to rethinking. Consequently, all of the truisms set forth by Einstein, to some extent, are now once again up for grabs.

For the most part, Tesla has been withheld from us. Why would this be? The reasons for suppressing his work could drive us into the heart of conspiracy theory, but it needn't do so. If we accept that commerce and capitalism are the real driving forces that really make the world "go 'round", we can see why Tesla and his inventions have been sequestered. But to get back squarely on track, Tesla's view was different than Einstein's; the *looking glass through which he sought to understand the Cosmos was electricity.* And this commitment was hardly for theoretical reasons – his theories resulted in real-world applications. How many know that he tested wireless power transmission? That he developed the basis for weather manipulation technology? That he invented the radio before Marconi? That he established a means to extract energy from the "medium" (aka the "aether") that would make purchasing energy unnecessary? Tesla was years ahead of his time. Maybe a whole century. And his inventions were extraordinarily dangerous – for many reasons.

ELECTRICITY NOT GRAVITY: THE COSMIC PRINCIPLE

What will not be revealed by the physics priesthood and those in control of them is the hidden essence of everything Tesla learned, employed and sought out for the free and mutual benefit of all humanity. To get to the point: electricity is foundational to what is. In other words, it's essential to the known universe. Its mechanisms are the functioning of electromagnetic forces. At the core of this awareness resides electric plasma, the so-called fourth state of matter, that has only really been understood in a *world made plain after the nuclear bomb*.

Why was Einstein's view adopted as orthodoxy while Tesla's work mostly rejected? To understand this, we must "go cosmic" and recognize that there is an author of deception, a father of lies, whom the Bible calls *Satan*. It is for his selfish purposes that he obfuscates truth and seeks to keep the world locked up in darkness. He has skilled minions everywhere (many of whom wear lab coats with pocket protectors). It is their assignment (most unwittingly of course) to employ the falsehood of gravity as the original essence of our universe for this surprising reason: Satan conceals his ultimate purpose not only to unseat the new "crown of creation" – humanity – but to destroy everything that God has made.[1] Satan seeks to annihilate God's creation.

You see, many academics put forth pronouncements concerning the universe based upon unbiblical postulates – i.e., the constant we call the

---

[1] Satan, oftentimes named Lucifer, was the most beautiful of all of God's creation (Isaiah 14). But his pride led him to sin. In aeons past, the archons, sometimes called archangels or "rulers of this world" (the taxonomy is not important to us here), while superior to human beings, fell from preeminence. Humanity was elevated in God's eyes and a new plan devised. And yet, while the archons still rule over nations and humankind behind the scenes, this structure has been officially revoked through the finished work of Jesus Christ (Colossians 2: 15). When Jesus Christ returns, the old creation will pass away, all things will become new.

speed of light, the singularity toward which the knowledge and intelligence supposedly evolves, and finally, the origin of all that is, the Big Bang theory. To date, despite the fact these concepts are largely adopted as unassailable axioms, they remain mostly unproven. New ideas, better ideas, are suppressed because the old ideas are accepted as dogma. Virtually every academic discipline abides by this suppression.

Figure 27 - BY FASTFISSION - FROM EDWARD GRANT, "CELESTIAL ORBS IN THE LATIN MIDDLE AGES"
ISIS, VOL. 78, NO. 2. (JUN., 1987), PP. 152-173.
Medieval concept of the Cosmos. The innermost spheres are the terrestrial spheres, while the outer are made of AETHER and contain the celestial bodies.

Satan energized the science built upon this false model of the Cosmos. His intention was to make humans look one direction – to distract humankind with a theory of the universe based on gravity – while he devised a plan to school a special cadre of practitioners on how electricity could be employed to create weapons of mass destruction. Satan thinks big – and he wants to destroy planets, suns, and galaxies. Other authors have taken up this subject elsewhere; Woodward discusses it briefly in this volume when he discusses the "real Star Wars". Author Joseph P. Farrell conjectures it has been accomplished before in our distant past by a "High Civilization" with technical knowhow far exceeding our own. Their knowledge led to a destructive event that ended the age and sent humanity back to the Stone Age (literally). However, for our purposes in this chapter, we need only make the point that in this aeon (in this age), God restrains Satan, holding him back from his official title as destroyer[2]. Satan must today work through humanity in accordance with certain rules. And Satan's program has been enhanced big time through the so-called "particle accelerator" in Geneva, Switzerland. We call it CERN. CERN is not one of a kind; but it is now the most visible and its scale way well beyond other devices similar in nature. CERN amounts to mechanizing Satan's *reason for being* (raison dêtre). Left unchecked, CERN can become that which Oppenheimer declared on July 16, 1945 when he saw the destructive force of the atomic bomb, "Now I am become Death, the destroyer of worlds" (from the *Bhagavad Gita*).

CERN is the largest machine ever built. Over a hundred nations participate in its financing. It takes only a modicum of discernment to realize its planners base their mission upon Satan's promise that *ye shall be as gods* after partaking of the *tree of knowledge*. For it's through this greatest of lies that humankind not only fails to detect its oncoming destruction but participates actively and eagerly in its self-annihilation.

TESLA TO THE RESCUE

Tesla held altruistic goals. He hoped to reveal the true nature of the Cosmos, and thereby free mankind from energy dependence of every

---

[2] Not only is Shiva, the goddess of destruction, *Apollyon* (or *Abaddon* in Hebrew), is also the God of destruction. The antichrist is associated with Apollyon (aka Apollo). "They had as king over them the angel of the Abyss, whose name in Hebrew is Abaddon and in Greek is Apollyon (that is, Destroyer). (Revelation 9:11, *New International Version*) We discuss this further in the following chapter.

sort. Tesla discovered how to tap into the electric plasma surrounding all objects in the known universe. He called this plasma, the "Aether" (derived from the Greek word for air or atmosphere, *aether*). In so doing, everyone would have free access to unlimited electrical energy.

But the masses have been kept in the dark (no pun intended). Access to the hallowed halls of science are off-limits to the lay person. The mainstream media mesmerizes the "herd" with graphical wizardry and fabricated authority amassed through repetitious broadcasts of televised pundits mouthing the same cosmic dogma again and again. Despite this obvious form of mind control – of technological deception – "the truth is out there" (as Fox Mulder from the *X-Files* would proclaim).

Figure 28 – SYMBOLS OF AN ALIEN SKY, THE THUNDERBOLTS PROJECT

What Tesla discovered to be true about the medium existent throughout the universe, Satan knew long ago. Through his sinister influence, the truth about the general availability of the plasma, the aether (now redefined in the new and ever-more popular theory of "the Electric Universe"), up to now has been bottled up. Resistant to change, popular physics continues to cite gravity as the explanation for why objects fall back to earth. Gravity lies at the center of the speculation about *Black holes*, *Dark matter*, and *Dark energy*. Physicists even invoke gravity as understood by modern physics to explain the expansion of the universe from the original "singularity" (not meant in the same sense as used by

Ray Kurzweil[3]), when all that now is, was at its genesis, condensed into an infinitely small ball of "everything".

So here we call into question the model the accepted academics continue to promote to the public. For if we fail to question those credentialed and persons of lofty positions, our enslavement will persist. Furthermore, left to their own devices (such as CERN, the biggest of all their devices), in their future – and our own – lies oblivion. Therefore, we must rise above the entrapment of force-fed perspectives and examine the universe afresh. Such requires more than merely shifting our attention, we must shift the paradigm. This is where Tesla comes to our rescue.

THE PARADIGM SHIFTS

The force of gravity compared to *electromagnetic* force, is extremely weak – ten-to-the-thirty-ninth power (10^{39}) weaker. When a baby holds up their rattle and shakes it, it proves the point readily enough. The electromagnetic force exists between electrically charged particles. Like charges repel, opposite charges attract. We all know this truism from about the Third Grade upward. We prove it so a thousand times a day.

Electromagnetism is the study of this physical interaction. The electromagnetic lines of force are detected as "fields". We learned about "force fields" in Star Trek. Force fields is "pop" vernacular for electro-magnetic fields. Instead of thinking about matter as energy compressed into a stable object of some sort, we need to rethink matter and energy as different forms of electricity. Matter and energy both vibrate – at different frequencies. Gravity is actually a weak form of electromagnetism, nothing more. Whether we examine the minutest particle (the quanta) at the tiniest end of the quantum spectrum, or the largest collection of particles

---

[3] Kurzweil uses the term singularity to be the point in the future when knowledge increases at a near-infinite rate. At that point, humankind will no longer be able to master knowledge as truth will be constantly changing. Discoveries will eclipse the old information continuously. Whether this spells doom or nirvana is a subject for debate. "While some futurists such as Ray Kurzweil maintain that human-computer fusion, or "cyborgization", is a plausible path to the singularity, most academic scholarship focuses on software-only intelligence as a more likely path." (https://en.wikipedia.org/wiki/Technological_singularity)

in stars, planets, and gases disbursed throughout a massive galaxy, electromagnetism behaves the same. As Trismegistus the mythical founder of alchemy, astrology, and magic supposedly said, "As above, so below". The rules hold true everywhere.

Plus, electromagnetic lines of force hold planets to orbital paths, not gravity. This is a big paradigm shift for us. As a force, gravity is too weak to account for the angular momentum[4] exhibited by planets expressed in accordance to their masses. Another major paradigm shift: *asserting that outer space exists totally empty or that it comprises a "vacuum"*. This is simply not true. Space consists of *electric plasma* throughout the Cosmos.

Plasma is one of the four fundamental states of matter. From our elementary school science, we learned that all matter exists in one of three forms: solid, liquid and gas. Today, we know *the fourth form is plasma.* Typically, a plasma is created by heating a gas, increasing or decreasing the number of electrons, which creates as a by-product either positively or negatively charged particles known as *ions*. Additionally, plasmas exist within intergalactic regions and can conduct (propagate) electricity. Keep this in mind as we consider the next paradigm shift, "gravitational pull at a distance". This truism has recently been shown *to be false.* Recently published papers in scientific journals demonstrate to the stubborn scientific community that neither gravity nor gravity waves are detectable. This means that, when taking into account the

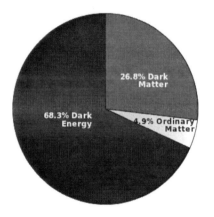

Figure 29 - DO WE REALLY DETECT JUST 4% OF THE UNIVERSE?

---

[4] *Angular momentum* for non-physicists is a difficult term. For our purposes here, we can express its meaning simply as the reason why planets keep spinning without slowing down and coming to a stop, and why they stay in place in their orbits without orbital decay until an external force (known as torque) influences them. Based upon an email discussion with friend Stan Deyo, Woodward noted from Deyo that technically, given enough time, the spinning movement of celestial bodies would slow and cease.

distances between planets (even within our relatively tiny solar system), *the weak interactions of gravity don't account for planetary orbits and their perturbations.*

Next, let's touch on *Dark energy* and *Dark matter*: These concepts are still topical on the *Discovery Channel* and the *Science Channel*. But they are nothing more than theoretical labels sold to the public at large. These concepts actually obfuscate the real dominance of electromagnetic forces and the existence of ubiquitous electric plasma. The commonly expressed view: most of what exists can't be observed or detected. 96.1% of what exists can only be "inferred". What we can detect amounts to less than 5% of the universe. We are literally blind when it comes to everything else. Does that seem reasonable? Does it seem like a stretch to believe?

Then there is today's favorite astronomical obsession, Black holes. Gravity-based models seek to account for the attraction of stars, planets and plasma in the formation of nebulae and other concentrations of energy and matter in the Cosmos. These remain elusive, however, as they can't be measured accurately nor be experimentally reproduced. We run across Black holes in virtually all discussions of space-time, warping of time, theorized wormholes, and multiple dimensions' dialogues, while actual observations, measurements and experimental laboratory-based models prove instead something very different: the existence of electrically charged plasmas which account for the behavior of matter and objects drawn to a centralized area of the Cosmos. The twin opposing *jets* of gamma and X-rays emanating from these so-called Black holes, are really the discharge of densely packed charged particles reaching a critical mass. These particles became tightly packed not due to the actions of gravity but due to magnetic attraction of opposing electrical charges.

THE GOLDEN AGE OF KRONOS – AN OCCULT MYTH

Building upon the discoveries of Tesla, recent work has been undertaken merging "plasma physics" with ancient anthropology – specifically analyzing paintings on cave walls supposedly crafted 20,000 years ago. Additionally, the written records of the Sumerians, Babylonians, Egyptians and Chinese seem to reinforce one another in astonishing ways. In fact, every culture including the indigenous natives of North America recorded heavenly events that testify to "the electric universe".

In short, what was observed and recorded globally, appears to be plasma discharges in the heavens surrounding our planet. And what is particularly intriguing and disconcerting at the same time, today's occult practitioners, masters of the dark arts, hold to an ancient story connected to the plasma physics exhibited worldwide. They affirm belief in what they label *The Golden Age of Kronos*. Being deceived by the lies of the Adversary, those illuminated by this mythical creed seek to return to this age of purported perfection before the dawn of our current civilization. Did it really exist? Was there a "High Civilization" that existed before our own? Was there an "Atlantis"? Their suppositions conflict with almost everything we know about our Solar System, its movements, and Earth as the sole source of life, the location where we believe God's greatest creation was fashioned.

Figure 30 - KRONOS (SATURN) WITH SICKLE

According to these myths, our Earth and Saturn were conjoined, moving through the Cosmos in a synchronized manner, enveloped within an atmosphere comprised of electrically charged plasma. Illuminating our planet was a *golden halo* – then Earth existed devoid of darkness. At that time, our planet travelled in tandem with other planets outside the boundaries of our present solar system. There was no Sun and thus no night. The electrified, golden plasma envelope provided illumination at all times. Indeed, it was a golden age in part because all that the creatures of this age observed was continuously imbued with a golden light.[5]

The return to this *Golden Age of Kronos* comprises the occulted mission of the highest, most exalted level of adherents. *Kronos* refers to Saturn as it provided the human inhabitants of Earth their only means to determine time itself.[6] Saturn – Kronos – was the perfect sun. According to its devotees, this mythos constitutes the ultimate truth about our Cosmos. In their view, our Earth and its place in the Solar System as we know it today, demonstrates how far away our planet and our race has slipped from the lofty civilization and cosmic structure of the previous aeon. With the exception of but a few traces in ancient literature, there exists no memory of just how advanced this civilization was. Although there are those who write about it with enthusiasm and sophistication (such as Joseph P. Farrell, Michael Cremo, David H. Childress, and a few others generally relegated to "fringe" science or "forbidden" history), the technologies employed during this time have been all but lost to humankind.[7]

Indeed, the nature of this supposed past has been kept away from the commoners comprising the 99%. The "keepers of this forbidden history", the 1%, are elites whose links to this halcyon time of yesteryear are but indiscernible whispers. Nevertheless, the implications resound loudly promising a strange, foreboding, and a most profane future predictably full of horrors, unbeknownst to its advocates but predictable to those tethered to the Bible. Characteristic of the science pursued in

---

[5] Some authors assert the Earth was illumined by a purple-hued light instead of a golden light. "At that time there was no Sun as we know it today. There was no way to tell day from night. No stars could be seen through the dense atmospheric purple haze and there was no moon from which to tell the passing of time by its phases or from which the Earth's oceans could be influenced in great tidal movements. Man lived in a perpetual state of dusky darkness. The warm and bountiful purple hue permeated all existence and the nocturnal thrived." McLachlan, Troy D. (2011-07-14). *The Saturn Death Cult* (Kindle Locations 166-169). Kindle Edition.

[6] The "time keeping" can be inferred, but it is complex and won't be taken up here.

[7] The quest to discover the secrets of this time long since passed, constitutes a major theme of the authors identified, and sometimes contains outrageous speculation that while entertaining and engaging, asks readers to suspend their disbelief to the extreme.

the name of returning to the Golden Age not only includes research at the CERN colossus, but laboratory investigations into genetic manipulation of human, animal and plant genetics around the globe. As in the days of Noah, so it is again today. Abominations abound. Chimeras become commonplace. Mary Shelly's *Frankenstein* foreshadows the evil from humankind's overreaching that "this way comes".

Was the "golden age" believed on by the most secret of secret societies mentioned at all in the Bible? Was this the world that God destroyed with water after humans practiced evil continually? Or was this so-called Golden Age from a different aeon altogether? Was it a time in which a "different man" lived among angelic beings?

The celebrated author of one of the most popular annotated study Bibles was the Rev. Finis J. Dake. In addition to the *Dake Study Bible*, he wrote a most underappreciated classic, *Another Time, Another Place, Another Man*. Dake argued that before Adam, the world was in fact a very different place. There existed a "pre-Adamite race" of humanoids that may have been quite distinct from our species. Could it be that it was this man from "another time, another place", and whose nature differed from our own, was the being who drew pictures on the walls of caves, composing drawings that reflected the strange sights in the heavens above? Dake references the famous passage in Ezekiel 28 that describes "the anointed Cherub" placed in the Garden of Eden and who held an exalted status, being Creation's more beautiful creature. Dake points out however,

> The Eden described by Ezekiel… was not the garden in which God placed Adam, for in Adam's Eden, "every precious stone" was not the covering of Satan, nor was he the perfect, sinless guardian cherub. Therefore, this must have been a pre-Adamite Eden, the garden that God made as the location of Lucifer's earthly kingdom, long before Adam walked the Eden of his day.[8]

---

[8] Dake, Finis (2011-01-06). *Another Time...Another Place...Another Man* (Kindle Locations 477-479). Dake Publishing, Inc. Kindle Edition.

In a second volume, we will discuss the nature of creation and delve deeper into the mythology of Kronos (we will touch on it briefly in the next chapter as well). Here, we will point out little more than that a connection exists between this legend and the mythical doctrine of gravity that science exploits to "keep us in the dark" to the true nature of the Cosmos, an understanding that promises so many benefits if it were disclosed. Instead, here our modest goal is only this: seeking to uncover that the deception portends far more than knowledge sequestered by the elite, merely for hoarding the mysteries of the universe for themselves, simply as a means to elevate their status in their own eyes, and potentially, in ours.

No, their sinister purpose goes much further. It involves "terraforming" the earth – remaking the planet into a home for a new species of human beings, made in the image of Lucifer, the *"anointed Cherub that covereth"*. (Ezekiel 28:14) It would literally be a "world of their own making" with a *"transhuman* race" whose total population would be carefully controlled to a "much more manageable number" of 500,000,000 (500 million) neo-human beings – meaning that by implication seven billion humans of today must be terminated by plague or other mass extinction events. In the following chapter, I will go deeper into the massive murder planned by the "powers that be" ... the manner and mechanisms by which Earth now undergoes terraforming on a planet-wide scale ... and how particle accelerators such as CERN play a major role in reshaping the earth, its oceans, and its tectonic behaviors.

However, to come back to the theme of this segment: we should recognize that humanity does possess a record of an era when overwhelming plasma discharges lit up the skies above. Of particular note, the patterns of these discharges were not random such as we witness in violent thunderstorms typical of the South and Southwest United States. No, the pattern of these discharges took on shapes that became familiar to its observers.

ANCIENT PHYSICS ON THE WALLS OF CAVES

The ancients recorded on cave and canyon walls, plasma discharges. These same human and animal-like figures they illustrated with primitive paints and paint brushes, have been reproduced successfully in the laboratory, including such renowned facilities as the Lawrence Livermore National Laboratory and the Lawrence Berkeley National Laboratory. (See the graphic below, retrieved from scholar David Talbott's website, https://ristorantemystica.files.wordpress.com/2011/11/plasma.jpg to whom we are all highly indebted for his noteworthy work).

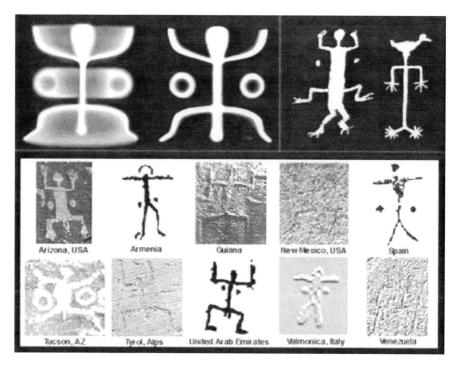

Figure 31- PICTOGRAPHS ON CAVE WALLS COMPARED TO PLASMA DISCHARGES IN THE LABORAORY

However, these correlations have not been brought to the mainstream. It has been left to the credentialed scholars to publicize it, such as theologian-cum-physicist Joseph P. Farrell and geophysicist-cum-anthropologist Robert M. Schoch.[9] Both discuss the research of astrophysicist

---

[9] While these individuals publish outside the "academically accepted journals" because their investigations and conclusions are non-orthodox, they are highly credentialed. Joseph P. Farrell has a Ph.D. from Oxford in Patristics (the study of the Church Fathers)

Anthony L. Perrault, who has worked for the Los Alamos Labs in New Mexico, who was one of the first to research and publish the fact that plasma displays in ancient times were recorded in caves across the globe. Schoch provides a summary of Perrault's findings. We include a small portion here:

> Peratt has studied ancient petroglyphs found around the world that are not only remarkably similar to one another, but also apparently record with a high degree of accuracy the configurations that intense electrical and high-energy plasma phenomena would form in the skies. He and his colleagues ... have recorded and analyzed the orientations of innumerable petroglyphs spread over 139 countries... The petroglyph locations generally have a distinct southern field-of-view, and Peratt and Yao concluded that the associated plasma phenomenon showed "relativistic electron flow inward at Earth's south polar axis and hypervelocity proton impacts around the north polar axis" (Peratt and Yao 2008).[10]

What the ancients etched into rock were the visions of electrical discharges occurring between Earth, Mars, Venus and Saturn. When in the proper proximity, these heavenly bodies linked and released electrically charged particles. These massive "electrical connections" were the plasma discharges scientists such as Schoch and Perrault credibly propose that gave rise to the pictures and petroglyphs observed worldwide. Remember: cave drawings are empirical facts. Their origins can be dated to some level of accuracy, and their patterns can be validated globally. We are not dealing with dreams or visions. (Talbott accomplished much of the research on the correlation between mythology and plasma discharges, providing this extensive empirical evidence).

---

while Robert M. Schoch is a full-time faculty member at the College of General Studies at Boston University since 1984. He earned his Ph.D. in geology and geophysics at Yale University. He also earned an M.S. and an M.Phil. in geology and geophysics from Yale as well as a B.S. in geology and a B.A. in anthropology from George Washington University. Anthony L. Peratt, is a plasma physicist with Los Alamos National Laboratory. His work is considered solid academically, but again, is not widely appreciated. The implications of his research challenge sacrosanct dogma. As stated elsewhere in this volume, orthodoxy suppresses all comers.

[10] Schoch, Robert M., Ph.D. (2012-08-10). *Forgotten Civilization: The Role of Solar Outbursts in Our Past and Future* (Kindle Locations 1585-1592). Inner Traditions Bear & Company. Kindle Edition.

Delving deeper, the mythos as articulated by a growing number of scientists enticed by the notion of the electric universe, proffers the following explanation: The *Golden Age of Kronos* existed before a major restructuring of the Solar System occurred, still within humankind's memory. This was known also as the *Era of the Gods*. Circa 3200 BC, Mars approached the ecliptic[11] at an angle of 24 degrees together with Earth and Saturn into our present Solar System. Interestingly, Earth's sky would not have radiated a golden glow. Rather, that of a *purple haze* – Saturn's outer egg-like plasma sheath, reflecting its dark light inwards producing a uniform dim purple glow on Earth – for Saturn was once a *brown dwarf star*.

Figure 32 - THE OCCULT SYMBOL OF THE BLACK SUN

It was the influence of the Sun's own electric field causing Saturn's plasma sheath to begin to wane. The ensuing reduction allowing for the replacement of Saturn as mankind's primary sun, with that of our present day Sun. This time period revealed to earthbound humans, a planetary alignment of Mars, Venus and Saturn. This is when they observed the many electric plasma discharges taking place between these planets and Earth; thus, spawning mythical stories of gods, constellations, perhaps even heavenly battles and wars. Until then, even the Sun itself remained blocked out by Saturn's plasma sheath.

In light of the implications of these findings, we might suppose that these studies would be acknowledged and publicized by mainstream science. However, such is not the case. Just as the discovery of giants

---

[11] Dictionary.com defines the *ecliptic* as "the great circle formed by the intersection of the plane of the earth's orbit with the celestial sphere; the apparent annual path of the sun in the heavens." See http://www.dictionary.com/browse/ecliptic.

across America have been cloaked by those fearful that such mind-boggling fossils would be he undoing of evolution, the standard view of the universe built upon its notion of gravity continues to suppress empirical data that would challenge its orthodoxy.

Perhaps it is only my supposition without documents that serve as bona fide smoking guns (for such evidence almost never surfaces being uncirculated among the initiated) – but one can readily assume that in an effort to maintain control over subordinates – monarchs, priesthoods, and secular dictators were established to withhold what was deemed the essential (albeit occulted) knowledge of how the universe operates. Information was obtained from "the gods". Satanic rituals opened portals across dimensional barriers, allowing entities through ... entities which possessed esoteric and arcane knowledge who eagerly conveyed such understanding to their human apprentices.

One such aspect of this forbidden practice was the *worship of the Black Sun*, as Saturn was known to its adepts. This Black Sun motif became an important part of Heinrich Himmler's SS[12] and remains a mosaic on Germany's Wewelsburg Castle floor in its north tower, and which may even today, serve as a location for Satanic rituals.[13] This worship carried with it supernatural power to manipulate the Cosmos. Passed down through the ages, this knowledge became entrusted to secret societies whose sole purpose was realizing a modern version of the Tower of Babel. One such society is *Fraternitus Saturni* (**The Brotherhood of Saturn)**, Germany's most secret and impactful secret society.[14] Today that tower has been built

---

[12] "The term *Black Sun* may originate with the mystical "Central Sun" in Helena Blavatsky's Theosophy. This invisible or burnt out Sun (Karl Maria Wiligut's *Santur* in Nazi mysticism) symbolizes an opposing force or pole. Emil Rüdiger, of Rudolf John Gorslebens *Edda-Gesellschaft* (Edda Society), claimed that a fight between the new and the old Suns was decided 330,000 years ago (Karl Maria Wiligut dates this 280,000 years ago), and that Santur had been the source of power of the Hyperboreans."

[13] Friend Russ Dizdar shared with Woodward his experience of seeing maidens in ceremonial attire come through the basement area at Wewelsburg Castle (where the symbol as shown on the opposite page lay) when he was conducting spiritual warfare there (coming against demons) he believes still inhabit the castle.

[14] According to Stephen E. Flowers, Ph.D., whose book, *Fire & Ice: The History, Structure, and Rituals of Germany's Most Influential Modern Magical Order: The Brotherhood of Saturn.* St. Paul: Llewellyn Publications (1994).

once again, a mechanism that can work wonders in connecting humankind to the gods, and whose self-admitted goal is to open portals to allow such lesser deities once again to transit to and from the Earth (and perhaps, to reshape and reform our planet). We know this modern Tower of Babel as the Large Hadron Collider (LHC), stretched across the French and Swiss border at Geneva, more commonly referred to as CERN.

Before we move on to the following chapter pertaining to this extravagant particle collider, let's dive a bit deeper into a biblical premise about the known Universe. God, our Creator, bestowed upon us minds limited to the boundaries He set forth. We are created in His image; so, we can obtain a sound understanding of the Universe He created. His mind far transcends ours. Thus, as intellectual Francis Schaeffer taught (in response to the dualism of Immanuel Kant), we can know things truly, just not exhaustively. There may very well have been a starting point, a singularity, where our Universe began. Scientists wish to promote this to we who reside outside their rarified priesthood. To them, science constitutes their only religion.

Beginning with a singularity, the simplest geometric form enclosing a volume is a tetrahedron. A polyhedron composed of 4 triangular faces, 6 straight edges, and 4 vertex corners. We conceive of it as a 3-D pyramid. In the beginning, Patch believes, this was the shape of our Universe in its vast expanse.

Likewise, if one were to examine at the quantum level the smallest arrangement of matter, you would observe a tetrahedral form. The large Hadron Collider at CERN lies in search of it. While its highest level minders won't admit as to its shape, residing within their own published journals, one discovers papers describing the C60 (Carbon 60) molecule formed of a cage-like structure – a regular truncated *icosahedron* – if you will, a soccer ball.

This same configuration is found in all interstellar matter. "As above, so below."

Building upon this singular citation, abundant support exists evincing the tetrahedron. For example, nano-scale diamonds, composed of C60 self-replicate, form a lattice framework akin to that of the original Buckminster "Fullerene" (see figure below). Such diamonds have been imparted with digital information using pulsed lasers, rendering data permanent in this storage medium. From there, a quantum computer, self-replicating outward into a spherical, 600-cell polytetrahedron, also known as a **C600**, hexacosichoron can be constructed. Several published manuscripts exist outlining experiments arranging nano-diamonds as well as *gold* into computational "qubits"[15] arranged as tetrahedrons. A quantum computer may be composed of such self-replicating nano-diamonds forming a spherical arrangement of tetrahedron qubits.

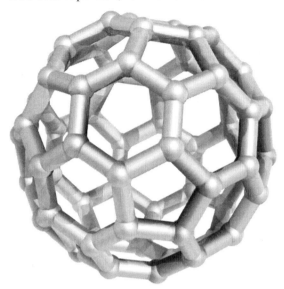

Figure 33 - THE FULLERENE OF BUCKMINSTER FULLER

---

[15] According to Wikipedia, "In quantum computing, a **qubit** (/ˈkjuːbɪt/) or **quantum bit** (sometimes **qbit**) is a unit of quantum information—the quantum analogue of the classical bit. A qubit is a two-state quantum-mechanical system, such as the polarization of a single photon: here the two states are vertical polarization and horizontal polarization. In a classical system, a bit would have to be in one state or the other. However, quantum mechanics allows the qubit to be in a superposition of both states at the same time, a property which is fundamental to quantum computing. A mnemonic for remembering the number of states of a single qubit is quad-bit or qubit."

Classical computers are based upon a single bit, a 1 or a 0, turned on or off. Quantum computers are based on two levels of information, or a bit that can have four states. It is a "cube bit" or "quad bit" or a qubit. This mimics particle physics in which particles can have full-spin states or only half-spin states. It is also known as "superdense coding" requiring far less storage to record information associated with such quantum information. Interestingly, the term's origin was a word play on the ancient form of measurement, the cubit!

They know. They know – but are not letting on as to the true geometry of our Universe. Unceasingly, *they* put forth their stories of the *Big Bang* in conjunction with their gravity-based model to explain the cosmological mechanisms to us, the unsuspecting masses. This is the false paradigm of our Universe. We have been misled into believing it.

The real model is founded upon electricity: *electromagnetism, not gravity*. The universe filled with electric plasma. It is a Cosmos designed at its largest vastness to its smallest particle... geometrically arranged in self-replicating tetrahedrons. The same is true whether examining the quantum scale singularity or, the cosmological galaxies. It is always the same shape. The tetrahedron is a cosmic principle. At its outer limits, Patch contends the Cosmos consists of a sphere – simultaneously elegant and extraordinary. What lies beyond only God can say.

CONCLUSION

Einstein did acknowledge Tesla as the *real* genius. Further, he acknowledged Tesla's goal of providing the world full and free access to the aether, the source of unlimited electrical energy. Unfortunately, only Einstein's earliest works involving gravity were addressed vigorously by his contemporaries. These resulted in the outright theft of Tesla's discoveries concerning electricity.

The obfuscation of the aether was necessary for profiteering derived from the generation and distribution of electricity. This established the pattern for all of modern scientific research and the buildup of the military-industrial complex which provides the destructive weaponry in all its raw power through which the elite rule the world.

What very few people know, and we hope you now appreciate this fact, is that the research and development taking place at CERN with its Large Hadron Collider, builds entirely upon the foundational work of Nikola Tesla. But while this may be news to you, rest assured *many of the more senior scientists in Geneva realize the true power of the universe that Tesla discovered.* Furthermore, many may reckon as Patch asserts, that the real geometry of this universe, its tetrahedron structure, lies coupled with the facts about electricity they keep under wraps. One can see this plainly enough by reviewing the configuration of the four main particle detectors within the LHC. *This configuration derives from tetrahedrons!* This is not a coincidence.

Within the following chapter, beginning with the present-day status of the LHC, Patch will propose the true and multiple uses of the largest, most complex, most expensive and (perhaps) most powerful machine ever conceived by man.

Then again… maybe such a machine had been created before and we just haven't yet learned where and when it was constructed. Perhaps that will be taken up in the next volume of our study. As the LORD leads.

THE GODS OF CERN AWAKEN
Anthony Patch with S. Douglas Woodward

And the nations were angry, and thy wrath is come, and the time of the dead, that they should be judged, and that thou shouldest give reward unto thy servants the prophets, and to the saints, and them that fear thy name, small and great; and shouldest destroy them which destroy the earth.

(Revelation 11:18)

THE FATE OF THE WHOLE WIDE WORLD HELD IN OUR HANDS

THE EUROPEAN ORGANIZATION FOR NUCLEAR RESEARCH (FRENCH: *ORGANISATION EUROPÉENNE POUR LA RECHERCHE NUCLÉAIRE*), ESTABLISHED IN 1954, OTHERWISE known as CERN, lies on the common border of the French and the Swiss. And despite what Al Gore thinks, this happens to be the real birthplace of the World Wide Web.

Today, as proclaimed by CERN's own press releases and statements, its leading researchers and scientists are seeking to open gateways to other dimensions... portals to other worlds. CERN's Director for Research and Scientific Computing briefed reporters stating: "Out of this door might come something, or we might send something through it." To quote other leading scientists at CERN: "...the Large Hadron Collider (LHC) the titanic machine may possibly create or discover previously unimagined scientific phenomena, or 'unknown unknowns' – for instance 'an extra dimension'".

The eminent physicist, Dr. Stephen Hawking has warned us that this machine could create such things as microscopic black holes. Additionally, in disrupting *the metastable vacuum of the universe*[1] which envelops the Earth, "disruptions" can come at us at the speed of light: *"This could happen at any time and we wouldn't see it coming"*. Hawking envisions a *vacuum bubble* moving at near the speed of light which would consume everything – the Earth and perhaps the entire Cosmos!

Warnings by other leading scientists outside research directly connected with CERN, speak of the production by the LHC of quantum-scale particles known as *strangelets.*[2] For our purposes, this "quirky" term means: "A quark-gluon condensate" resulting from particle collisions yielding energies *above a threshold level of 10 TeV*. (**TeV**, or Tera Electronvolt, is a measure of energy: 1 TeV equals one **trillion** electronvolts, or $1.602 \times 10-7$ Joules).

To date (since reaching **6.5 TeV** per beam on May 20, 2015 – times two beams – equals a total of **13 TeV** – therefore exceeding 10 TeV), **the particle collider has been producing** *strangelets.* Is there any reason we should be concerned about the creation of these new quirky (strange) miniscule quarks (particles)? I think there is, besides the fact that the officials at CERN denied that they would generate strangelets. Let's consider the unique characteristics of a *strangelet* whose genesis lies just outside Geneva.

---

[1] "In general a false or metastable vacuum is a local, but not global minimum of a potential. This means that a different configuration of a system, let's boldly say the Universe, would correspond to a smaller energy level. One of the principles of physics is that a system always tries to be in the configuration/state of the smallest possible energy." From Philipp Grothaus. See https://www.quora.com/What-are-metastability-and-false-vacuums-especially-those-of-the-universe-In-mostly-simple-words-please.

[2] According to Wikipedia: "A *strangelet* is a hypothetical particle consisting of a bound state of roughly equal numbers of up, down, and **strange quarks**. Its size would be a minimum of a few femtometers across (with the mass of a light nucleus). Once the size becomes macroscopic (on the order of meters across), such an object is usually called a quark star or *"strange star" rather than a strangelet*. An equivalent description is that a *strangelet is a small fragment of strange matter*. The term "strangelet" originates with Edward Farhi and R. L. Jaffe". See https://en.wikipedia.org/wiki/Strangelet.

JUST HOW STRANGE IS A STRANGELET?

First off, the *strangelet* is the **most powerful explosive substance in our known Universe.** Yes, that's correct: THE MOST POWERFUL EXPLOSIVE SUBSTANCE ANYWHERE. Secondly, given its amazing density, **it cannot be contained** (it can't be put somewhere and retrieved later – i.e., to be uncontainable is to be unrestrainable). To make my point perfectly clear: what we can't "put in a box" (so-to-speak), we can't control. Thirdly, **its charge state is stable** relative to ordinary matter which by definition is *unstable* (however, counter-intuitive that may seem). Because strange matter starts off stable but ordinary matter is not, the two are destined for what could become *a fatal attraction.* In other words: *strangelets* attract ordinary matter to themselves. Once these particles come to life... once they are produced (which is what happens during particle collisions of protons and/or ions of lead as "manufactured" at the LHC), since *strangelets aren't neutral* – i.e., they aren't *unattractive* (like their historically politically *neutral* hosts, the Swiss – figuratively speaking of course), **they plummet to the core of our planet.** Upon arrival there, they come to rest due to the balancing effect at the earth's core, i.e., being at the center of our planet's electromagnetic field. But they aren't really resting once they *journey to the center of the Earth.*[3]

Indeed, once they are parked there, all is not right in our world's innermost "chamber" as we might wish. Many suggest the effects of strangelets are actually felt by our planet. The surrounding ordinary matter itself (which makes up our Earth of course) *being unstable* in its charged state, is drawn toward the stable *strangelets*. These potentially detrimental "lures" manifest themselves back near the surface of the planet in the form **of tectonic movements and volcanic activities**

---

[3] Are strangelets dangerous? To review why CERN believes they are not, this article addresses the fears asserted by Hawking and other scientists. See http://press.cern/backgrounders/safety-lhc. Is this propaganda or truth? It is the case of taking one expert's opinion over another's. CERN certainly wants to convey it is safe. Others express strong doubts. Be mindful that CERN denied they would ever create strangelets. *And they did create strangelets.* Now they are creating massive amounts of strangelets. And they don't know what they don't know.

along Earth's crustal plates. This sounds really dangerous. But hold your breath because it gets worse.

Depending upon the quantities of *strangelets* produced, within anywhere from as short a time as 10 years to ten times that many years (that's 100 years once you do the math), Earth may be converted to a **neutron star** composed entirely of this densely compacted *quark-gluon condensate*. I can promise you that this would not be a conversion we would want to witness firsthand.

In this year of 2016, the LHC at Geneva will produce many more *strangelets*. But this is not the lead story. A new particle accelerator known as *The AWAKE Experiment* sent its first beam down its corridor on 17 June 2016 within the vast confines of CERN. This is an electric, plasma-based, *linear accelerator*.[4] The Advanced Wakefield Experiment (AWAKE) will yield energy levels at least 100 times those of the LHC itself. That would be 1,000 TeV, or 1 PeV (Peta Electronvolt or one **quadrillion** volts). The promise of this approach to particle acceleration constitutes drastically reducing the area needed for particle accelerating, potentially making it all the way down to the desktop.

Wow! Think about that. First, we had a personal computer for your office. Secondly, we now have 3D printers able to manufacture objects at your home. And finally, we might soon own a particle accelerator *for your desktop!* The only drawback to that wonder: once the masses have their personal particle accelerator, we could participate as a member of say, the Shiva Shake-up Club, generating massive earthquakes, and *join as one in the destruction of the planet.* While it would be a marvel of progress, obviously it's not the kind of progress we should pursue for the sake of humankind.

At first glance, one might surmise AWAKE to overtake the LHC, rendering it obsolete. However, research and development is well underway in the construction of ever more powerful superconducting magnets

---

[4] A **linear particle accelerator** (often shortened to **LINAC**) is a type of particle accelerator that greatly increases the kinetic energy of charged subatomic particles or ions by subjecting the charged particles to a series of oscillating electric potentials along a linear beamline." See https://en.wikipedia.org/wiki/Linear_particle_accelerator.

for the ring-based LHC accelerator. This will extend the life of the world's most expensive machine, now expected to remain serviceable until 2030.

The LHC and AWAKE particle accelerators will soon, literally, join forces. Given what seems to be a cosmic catastrophe just waiting to happen, we are wont to ask, "What are the mad scientists at CERN seeking to accomplish by doubling down and combining the accelerators? As mentioned earlier, the masterminds are hunting for *additional dimensions and the means to access them*. Yes, as stated at the outset, they hope to open a portal. The question is, "A portal to what? For what reason?"

Figure 34 - THE CERN LOGO - THE CIRCULAR AND LINEAR ACCELERATORS COMBINED? A PATTERN OF ANTICHRIST'S 666?

The challenge before them constitutes the breaking the strongest of four fundamental forces in the universe: it is known as the *strong force*. (You might recall that the three other forces of the so-called Standard Model of physics are the (1) *weak* force, (2) the *electromagnetic* force, and last and almost least, (3) *gravity*.) A quantum particle – a boson known as the *"gluon"* – carries or conveys this *strong* fourth force.

Remember the gluon? We mentioned it when we introduced *strangelets*, aka a **quark-gluon condensate.** Perhaps we should advance our meaning by asking a question: What do you have when you have a whole lot of strangelets? You have a *bunch of quarks* stuck together on a massively small scale that – when they get together – can blow up your planet and really ruin everyone's day. When you see the statue of Shiva lurking in CERN's courtyard (the God of destruction), it makes you wonder if the scientists at CERN have a shared death wish for humanity. The fact is that such a death wish is really not so far from the truth.

Now you can begin to see why creating massive amounts of strangelets is not such a bright idea. We ought to wonder why dozens of nations came together and spent billions developing a brand new way to maybe, just maybe, blow ourselves up. Later in this chapter, I'll speculate for you why I believe countries that are otherwise political enemies have been willing to join forces to accomplish "the Great Work", but what that work will likely achieve, much to the detriment of humanity.[5]

BREAKING UP IS SO HARD TO DO

By virtue of its 27-kilometer configuration, the LHC will produce a *toroidal field*. What is that you ask? It is a doughnut-shaped field resulting from the electromagnetic lines of force generated by both its superconducting magnets, and accelerated protons (held in line by those magnets) to follow circular paths within the cryogenically cooled vacuum of *"the Pipes"* (as the 27-kilometer long main ring of the LHC is also known). Now we all know Switzerland is a cold place. But the pipes of the LHC are colder – much colder. For inside *The Pipes,* the interior stays colder than most temperatures in outer space.[6]

Anticipating you are wondering just how cold outer space is, allow me to digress. To begin, it varies based upon *whose asking* (or more accurately *what is nearby* and whether more photons are being absorbed or being emitted – losing photons makes you colder while gaining photons makes you feel warmer). *Absolute zero* constitutes approximately *a minus 270 Celsius.* Pluto, our distant micro-planet friend, is almost

---

[5] *The Great Work* was identified by *Eliphas Levi* (1810-1875) with the following definition: "The Great Work is, before all things, the creation of man by himself, that is to say, the full and entire conquest of his faculties and his future; it is especially the perfect emancipation of his will." This was transformed by Aleister Crowley into the *Law of Thelema.* In essence, it is the "triumph of the will" as Nietzsche would define it for the Ubermensch. *Triumph of the Will,* was a documentary without narration by filmmaker Leni Riefenstahl concerning six days of Hitler's activity in 1935 at Nuremburg.

[6] *Synchrotron Energies* should be mentioned in passing. They are Gamma and X-rays produced by the circulating particles within The Pipes, accelerated to 99.993% the speed of light, mimicking naturally occurring gamma and X-rays in our Cosmos.

that cold. In other words, being on Pluto would be just as frigid as being inside the pipes of the LHC![7]

But getting back to physics. The AWAKE linear particle accelerator, rather than employing magnets to force particles to scurry down a curved pathway, instead moves them in a straight-line configuration to near the speed of light, utilizing waves of electrically charged plasma and a pulsed laser. Electrons are used as a "witness beam" (think tracer bullets fired from a machine gun), while protons (and nearly any other type of charged particle) may be accelerated by employing this machine's esoteric services.

Before too much longer, however, both the AWAKE and LHC accelerators will be combined, projecting a linear beam of charged particles through the center of the circular toroidal field. This beam will be focused and directed by the toroidal field, targeting a very specific point (if you want to know where it may be pointing and why, keep reading). It is at this moment when the *strong force* will be broken, with a resulting gap created between the quantum particles. Then a portal will have been pulled out of thin air, ripped open if you will, creating a doorway to another dimension. In this instance, the "gap" as a pathway to God takes on a profound new meaning from what theologians historically have discussed.[8]

---

[7] "The surface temperature of Pluto can get as low as -240 Celsius, just 33 degrees above absolute zero. Clouds of gas and dust between the stars within our galaxy are only 10 to 20 degrees above absolute zero. And if you travel out far away from everything in the Universe, you can never get lower than a minimum of just 2.7 Kelvin or -270.45 Celsius. This is the temperature of the cosmic microwave background radiation, which permeates the entire Universe. In space? It's as cold as it can get." See http://www.universetoday.com/77070 /how-cold-is-space/.

[8] Woodward notes, theologians in the twentieth century wrote about "the God of the gaps" as a criticism of theological thinking which postulated God exists in the spaces where we lack knowledge. The idea: God exists as the necessary means to make things work that we don't understand. Therefore, God resides in mystery. However, as humanity advances to learn more and decode the mysteries, God grows smaller. The cliché, "The devil's in the details" is a corruption of this idea. Proposing the creation of a physical gap in physical reality by breaking the strong bond to open a way to "God" infers a *path to God* that would marry the mystical and the physical. But it may be entering a backdoor possibly seen as intruders by the entities who reside there. What lies beyond on the other side of the doorway? We may be on the threshold of finding out.

THE GODS AROUND (AND UNDER) THE MACHINE

We should not overlook the point that the LHC is located on the site of *the ancient Roman temple to the Roman god, Apollo*. This is no accident. Straddling the border between France and Switzerland, CERN extends through the French town of *Saint-Genis-Pouilly*. The name *Pouilly* derives from the Latin "Appolliacum". In Roman times, the site was believed to be a gateway to the Abyss. Given that I believe the planners of the LHC had an occult intent in mind when they devised and built the Machine of machines, it takes on special spiritual significance.

Now that I have your attention, let's ask together, "Why, pray tell, are men and women of science delving into matters of spirituality? Why do they want to connect to pagan, ancient gods, while refusing to acknowledge the real Creator of our Universe?" Permit me to place this question into a biblical context. Look at Revelation 9:1-11 (quoting here from the New International Version – NIV):

> *¹ The fifth angel sounded his trumpet, and I saw a star that had fallen from the sky to the earth. The star was given the key to the shaft of the Abyss.*
>
> *² When he opened the Abyss, smoke rose from it like the smoke from a gigantic furnace. The sun and sky were darkened by the smoke from the Abyss.*
>
> *³ And out of the smoke locusts came down on the earth and were given power like that of scorpions of the earth.*
>
> *⁴ They were told not to harm the grass of the earth or any plant or tree, but only those people who did not have the seal of God on their foreheads.*
>
> *⁵ They were not allowed to kill them but only to torture them for five months. And the agony they suffered was like that of the sting of a scorpion when it strikes.*
>
> *⁶ During those days people will seek death but will not find it; they will long to die, but death will elude them.*
>
> *⁷ The locusts looked like horses prepared for battle. On their heads they wore something like crowns of gold, and their faces resembled human faces.*
>
> *⁸ Their hair was like women's hair, and their teeth were like lions' teeth.*

> *[9] They had breastplates like breastplates of iron, and the sound of their wings was like the thundering of many horses and chariots rushing into battle.*
>
> *[10] They had tails with stingers, like scorpions, and in their tails they had power to torment people for five months.*
>
> *[11] They had as king over them the angel of the Abyss, whose name in Hebrew is Abaddon and in Greek is Apollyon (that is, Destroyer).*

Apollyon is Apollo. And guess what. Apollo is not the only god to whom the devisers of CERN pay homage. Recall the picture in the last chapter of the Hindu god named Shiva right in the middle of the grounds. Okay, you now ask, "Why did CERN's founders decide to place a statue of the Hindu god Shiva outside the headquarters building?" Given that Shiva is a god worshipped for his powers of destruction – laying bare the Cosmos (followed by rebirth), what meaning does this imply? Note: Christianity lies wholly devoid of any mention by the minders of the LHC, much less the placement of a cross upon the "hallowed" grounds at CERN (hallowed that is, to pagan gods). However, I don't mean to sound surprised by their choice. It is not a simple oversight. It is most intentional. **It has been my contention all along that the ultimate objective of CERN concerns spirits, not just physics.** Not only that, to be much more specific, I believe CERN quests to birth a new race of beings. To accomplish this great work, it will enlist support from agents of the dark side. That is, *the LHC intends to play a part in re-creating humanity and fulfilling "the Lie" as propagated by Satan himself.*

We know humanity was promised divinity by Lucifer a long time ago. However, here we are once again, having come full circle (albeit a ringed particle accelerator circle in Switzerland) to devise yet another way to become as gods. No doubt the scientists at CERN, almost all of them I think, are unaware of any sort of Luciferian agenda – just like Freemasons are unaware of the Illuminist adepts at the heart of their ideology. Nevertheless, I believe this reveals what the fuss at CERN is really all about. Yes, the true story will sound exactly like a Dan Brown novel, and prominent aspects of many other science fiction stories to boot. But hold on to your hats – we are just getting started.

THE HIDDEN AGENDA BEHIND CERN'S COLLISIONS

Displayed behind curved glass panels are ancient parchments and animal skins which contain formulae for communing with the gods. Many paranormal practitioners derive forbidden knowledge from corrupted deities and demonic entities Such things as long-hidden incantations include instructions describing (for example) frequencies of audible sound energy that must be produced to open dimensional portals. Another amazing matter that could be asked in the form of a question is this, "How does one communicate with spirit entities existing 'on the other side?'" Because it's timely I will drop a hint here: the Tower of Babel was (more than likely) attempting the very same thing.

A "Dance of Symmetry" film was produced in 2015 by CERN, replete with ceremonial dance movements meant for both the inducement of trance-like states in the dancers, as well as vibrations and frequencies tuned for communicating with an uncharted dimension. Once again, such inexplicable activity begs the questions, "Why is the scientific community involving itself in what (to the outside observer) appears most unscientific? Is it not a ceremonial, if not religious ritual?" The practice has been repeated again in Switzerland on June 1, 2016, at the opening of the Gotthard Base Tunnel. Was it a satanic ritual? No doubt it was a grossly pagan one with leaders of Europe prominent in their attendance.[9] As Stephen E. Flowers states in his book, *Fire and Ice*, Switzerland, Austria, and Germany are home base for occultist practices.[10] (England, it should be noted, wasn't far behind.) Coincidence?

These types of questions have been posed to the leadership of CERN. In the case of the glass-enclosed artifacts, the puzzling placement of these items at CERN is dismissed as simply a reverence for our ancient past or perhaps an artistically choreographed celebration of the ongoing achievements by the brilliant minds among the community of scientists and engineers at CERN. It is harmless: a self-indulgent

---

[9] See http://www.inquisitr.com/3165856/satanic-ritual-in-switzerland-worlds-longest-tunnel-opens-with-bizarre-ceremony-video/ for details and videos.

[10] See Flowers, op. cit., pp. 7-14.

recognition of human alchemists whether magician or technologist. Of course, one such achievement takes note of their purported discovery of the *Higgs boson particle*. Coined by the press as the "God Particle", it is considered one of the first particles produced during the earliest moments of the Big Bang (*earliest moments* as in a split-second). To be sure, this was a remarkable story of what humanity can do when combining its efforts to achieve something extraordinary. If the motivation was only purely research for the sake of science and achieving indisputable benefits for humanity, it would all be well and good. Unhappily, there's much more to it than that – *the agenda is sinister.*

Additionally, as a Christian, I continue to be struck by the outright dismissal of any consideration of the Bible and its relevant message as well as praise to the Divine Author and Creator. The top minds revere their own mental capacity, but not the mind of the One who created the Cosmos "in the beginning". This oversight (and I deem it quite intentional) is made even more breathtaking when considering the occulted behavior of the operators of the various ultra-high-tech devices submerged three hundred feet beneath the surface of France and Switzerland. The outright hubris of the promoters for the multi-faceted hidden agenda must not be overlooked, especially by those of us who humbly revere the God of the Cosmos who deserves far more praise than those who only discover His ways.

Indeed, their plan amounts to much more than simply discovering the particles composing the physical reality we call "the Universe". Likewise, it involves much more than opening dimensional doorways.

The *Masters of the Machine* seek an ancient promise, that *"ye shall be gods, having obtained the knowledge of good and evil"* (Genesis 3:5). What is being reaffirmed in humanity's latter days – just as it was during its early beginnings – is that humankind, through the exalted acquisition of pristine ancient knowledge (*prisca sapientia*[11]) has *chosen evil over good.* Perhaps this should be expected, but it remains regrettable.

For the self-proclaimed "lofty" men and women of science, in *"professing themselves wise, they became fools"* (Romans 1:22). They have chosen consciously and deliberately to believe "the Lie". In their unconcealed overconfidence, the truth is that they strive for *immortality* through the prowess of scientific research and its derivative machines. On the surface this seems like an outrageous allegation. But I will explain why CERN plays an important role in this occulted quest. With God's leading – accompanied by your patience – it should become clear to most readers why *CERN has become the modern Tower of Babel, and its leaders the ideological sons of Nimrod.*

A DNA MAKEOVER MASQUERADING BEHIND THE LHC

The impetus for building the Large Hadron Collider today is no different than it was for the Tower of Babel long, long ago. Many teachers believe that in the days of Noah, the fallen angels and perhaps their offspring, the Nephilim, engaged in manipulating human DNA. The implications for unsaved humanity, its vainglorious destiny, will be the same as Noah's generation spread across the face of the antediluvian world.

> *And God saw that the wickedness of man was great in the earth, and that every imagination of the thoughts of his heart was only evil continually.* (Genesis 6:5)
>
> *As it was in the days of Noah, so it will be at the coming of the Son of Man.* (Matthew 24:37)

We uncover this pernicious pursuit by diving deep into the origins of particle accelerators. We discover there that our human DNA and RNA are the very things that Satan intends to alter. He hates all of

---

[11] "It's ironic that most of the men who participated in the "scientific revolution", whose contributions seem (to us) so original and innovative, were themselves convinced that they were merely re-discovering the vast body of pristine knowledge (*prisca sapientia*) that had been possessed by the ancients, but somehow lost and forgotten during the centuries that came to be called the 'dark ages' of western civilization." See http://www.mathpages.com/home/kmath066/kmath066.htm.

God's creation. He especially hates man and woman, as we are image-bearers of our Creator. And yet, he seeks our worship and he desires that by changing our nature, *we will reflect his image instead of Yahweh's*. His goal guides and undergirds *Transhumanism*. His mission amounts to nothing less than the creation of the Übermensch of Nietzsche.[12]

Science history credits Dr. Ernest Lawrence with the invention of the circular type of particle accelerator. His machines continue operating at the Lawrence Livermore National Laboratory and the Lawrence Berkeley National Laboratory, as well as hundreds of others proliferated around the world. He began with small handheld prototypes, which evolved into the Bevatron, then cyclotron, and on into our day with the Synchrotron housed within the Advanced Light Source Building on the U.C. Berkeley campus.

As a pre-teen, this author accompanied his physician father, to visit the early machines on the Berkeley campus that I would later attend as an undergraduate. At that time, it was also the formative period of oncology – in particular, oncology addressing the thyroid. Dr. Lawrence's machines bombarded cancer in the thyroid using particles accelerated to high energies. Patients were transported "up on the hill" to the "Rad Lab" overlooking both the campus and the San Francisco Bay. There they sat passively on a chair, alongside a small sliding panel separating them from the massive magnets which were imposingly housed within that circular building. Had any of the patients actually peered at the machine on the other side of this panel, they might have had second thoughts about submitting their sick thyroid to the target practice of this machine and its (presumably) benevolent minders. Of course, these unfortunate folk were patients suffering the end-stages of cancer. They

---

[12] "Despite its growing popularity, many people around the world still don't know what 'transhuman' means. *Transhuman* literally means beyond human. Transhumanists consist of life extensionists, techno-optimists, Singularitarians, biohackers, roboticists, AI proponents, and futurists who embrace radical science and technology to improve the human condition. The most important aim for many transhumanists is to **overcome human mortality, a goal some believe is achievable by 2045.**" (Emphasis added. From http://www.huffingtonpost.com/zoltan-istvan/a-new-generation-of-trans_b_4921319.html).

placed their scant amount of hope in medicine's latest miracle. Understandably, they were willing subjects with no other alternative.

But today, the state-of-the-art machine is a *Synchrotron*, producing the most powerful X-rays outside of the LHC at CERN. Thus, the facility bears the name the Advanced Light Source Building (ALS). Perhaps surprisingly, *it is here the human genome project originated.*

This is so because X-rays "infer" the location, size, shape and arrangement of DNA and RNA. *Successful sequencing of human DNA is an outgrowth of particle accelerators.* While three-dimensional computer modelling of the amino acids and proteins took place within a computer, they were reconstructed outside it, in a laboratory.

Today, beginning at the quantum scale (at the tiniest end of the physical spectrum) these actual particles, smaller than atoms, are sequentially arranged *according to the designs of man* – and not of God. So it is we have entered into a new "mini era" of the longer-lived so-called "Digital Age". We may call this period the "era of *digitized DNA*". It is fast becoming the most dangerous time in human history.

How does it work? Biological sampling is conducted at one location, with the sequencing of DNA at another. First, digitizing the twin helical structure occurs. Then scientists transmit this digital coding to another sequencer, which reassembles the original patterns of DNA.

Remember, DNA is the acronym for *Deoxyribonucleic Acid*. Four types of nitrogen bases compose these nucleotides: adenine (A), thymine (T), guanine (G) and cytosine (C). These are chemical codes whose combination lies beneath every creature on earth, whether flora or fauna – planet or animal – encoding every single distinctive feature of those life forms. Whatever "coding" humanity has engineered to replicate realities, it pales in comparison *to the coding and the miniaturization designs God devised in creating us.* (Of course, the evolutionist argues such incredible design work was accomplished purely by chance).

Apart from the original biological sampling site, for the first time in the recorded history of humankind (notwithstanding an advanced ancient civilization whose existence is hinted at by what author Michael Cremo calls "forbidden archeology", but hidden by the mists of time), the reassembly or re-sequencing of DNA carries with it *the* most serious implication. For not only can the original pattern of DNA be replicated, it can also be altered either prior to or upon receipt of its transmission. Like packets of information carried over the internet, DNA (akin to packets) can be received in a sequence that is *"in* or *out* of order". Routers (which we all utilize at home to connect our computer to other computers in our house or to the internet), have a means through which they verify a sequence is correct and the packets are represented in the right structure (called "parity checking").[13] However, genetic scientists are the "routers" of the DNA sequence. And parity checking isn't in play when humans are manipulating or reordering the DNA "packets". This enables the alteration of DNA, effectively *mutating* the original. Technically speaking, the resulting entity is altered (or constitutes a *chimera* or a *hybrid* if the sequence is borrowed from another creature's DNA).

New forms of biological life can and have been realized in the laboratory (whether "mutants" or "hybrids"). We must ask, however, "Has *life* itself been *created*?" In my opinion, "No". Only our Creator gives the gift of life "from scratch" (or as the theologians would say, *ex nihilo*, "from nothing" – see Hebrews 11:3). However, actual living organisms have indeed been manufactured from that which already exists. Beyond the "Petri dish level", we now know animal and even human organs have been cultivated at labs globally. So it does not take a great leap of faith to conclude entire bodies of creatures, even human bodies, have possibly (if not probably) been fabricated already as well.

Thus, we have the *imminent arising of human hybrids* soon to be released into our world (some argue they already "walk among us.") What was once the stuff of ancient myths or modern science fiction, especially the accounts of Nephilim as discussed earlier in this book, now overtakes us. These offspring of the fallen "sons of God" mating with the "daughters of men" in Genesis 6:4, (or the giants inhabiting Canaan, e.g., Numbers 13:33), now more than ever before, deserve our

---

[13] *Parity checking* is a mathematical algorithm to validate the sequence of packets in a packet-switching network.

attention. Likewise, the chimera hinted at in the Hebraic apocrypha are "real-time" sketches of what happens today in unbridled genetic labs.

"And God said, Let the earth bring forth the living creature after his kind, cattle, and creeping thing, and beast of the earth after his kind: and it was so". (Genesis 1:24) The boundaries of what God intended have been breached by humanity unfettered by God's loving restraints. The horrors of what lies ahead will exceed our wildest imaginings. And CERN plays no small part in the debut at this horror show's premiere.

THE OPEN DOOR POLICY AT CERN

The most serious question is this, *"What exactly will pass through this portal* the LHC at CERN opens?" Dr. Stephen Hawking warned us in a documentary for the Discovery Channel: "If aliens visit us, the outcome would be much as when Columbus landed in America, which didn't turn out well for the Native Americans." He argued that instead of trying to find and communicate with life in the Cosmos, humans would be better off doing everything they can to *avoid* contact. It always seems to me a curiosity that human beings think *extraterrestrials are benevolent,* especially when the activity reported is often *malevolent.*[14] And yet, no one else seems to take his caution seriously. So why then are the majority within the scientific priesthood ignoring Hawking and pushing forward at the LHC to realize a literal breakthrough – to tear a hole in the fabric of reality – *opening a door* for what may be enemies of our earthly lives if not our very souls?

Ultimately, the reason constitutes *deception*, pure and simple. Lies have been put forth by the father of lies and scientists accept the lie of Lucifer (whether wittingly or unwittingly they accept what comprises

---

[14] Author Jeff Wingo offers this informed opinion: "If aliens are real, and if they have been visiting the Earth for years as many people claim, have you ever considered the morality of their actions? If you have, then you may have come to the same conclusions that I have, or maybe not; it all depends on your view of the alien and UFO phenomena. By that I mean, are they extraterrestrial beings or do they represent the nefarious presence of something else that is indigenous to the Earth?" From https://alienantichrist.net/2015/02/28/alien-visitation-history-tells-the-truth-by-jeff-wingo/

the premise of Luciferianism – that *knowledge equals salvation*).[15] These are but mere mortals seeking the reality of immortality. They seek to be as gods, rejecting that they are beings created in the image of God Himself. Satan stalks them promising a Transhumanist future – a transformation into godhood they he asserts comes through the knowledge of changing our genetic makeup.

Now follow me closely.

Christians are prone to jump to the conclusion that what will come through the portal amounts to wicked spiritual beings: demons or spirits, perhaps angelic in nature, manifesting in physical form once permitted through the portal. Is jumping to this conclusion warranted? Maybe not.

We must be cautious in quickly adopting an image of a spirit being or just its DNA coding sequence, however, squirting through the miniscule opening. We have entered into a realm where physical and spiritual are not so easily distinguished. Furthermore, the distinction between "atoms and bits", between physical reality and digital reality, also becomes blurred. This realm is not native to us – it stretches our comprehension and our creeds. It requires *revising reality;* that is say, we must rethink our worldview.

In fact, I believe that what is to pass from one realm into our own will be *transmitted in digital form* (perhaps the better word would be *transfused like blood* – the elixir of life) as what comes through isn't really just a sequence of numbers. What secretes through the portal, digitally, is not the commonplace binary coding used by computers "up to now". No. The construct of what comes through the door at CERN will be *quadnary* – a term I coined in my two novels.[16] Recall the definition supplied at the conclusions of the last chapter in a footnote:

---

[15] Jesus said, *"Ye shall know the truth and the truth shall set you free."* (John 8:32) He did not say, "Ye shall obtain knowledge and it shall be your salvation" (essentially the core of Gnosticism) But what is the *truth* to which Jesus referred? *"I am the way, the truth, and the life, no one cometh unto the Father but by me."* (John 14:6)

[16] Classical computing uses binary, or two bits to represent two states as "on" or "off". Quadnary implies four bits, quad-nary instead of bi-nary, to represent these four states, mimicking as we said, the spin states of particles. Perhaps it is no coincidence that DNA uses *four* nucleic acids whose combinations encode all of life itself. Incidentally, my

Classical computers are based upon a single bit, a 1 or a 0, turned on or off. Quantum computers are based on two levels of information, or a bit that can have four states. It is a "cube bit" or "quad bit" or a qubit. This mimics particle physics in which particles can have full-spin states or only half-spin states.

 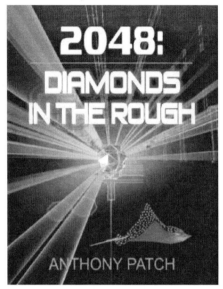

Figure 35 – NOVELS WRITTEN BY ANTHONY PATCH

Such is the construct of a quantum computer comprised of qubits, rather than transistors. This is an entirely new form of computing. It may also comprise the structure of an altogether *newly created beast*. It may consist of, or be the very essence of the *Beast to come* as described in Revelation 13:1-18. (Gonzo Shimura will talk about how a "networked Chimera" may arise in a later chapter of this book, perhaps based upon quadnary encoding).

This computer, of which already many are in operation at such places as JPL, Lockheed, and NASA, are described as the *Adiabatic Quantum Computer* (AQC). In its most recent iteration, its manufacturer proclaims it in possession of processing power equivalent to that of over

---

two novels are *Covert Catastrophe* and *2048: Diamonds in the Rough*. Both deal with these subjects. Check out my website to learn more about them...anthonypatch.com.

seven billion human brains.[17] Additionally, it derives solutions to combinatorial problems from other dimensions. Literally operating at the quantum level of physical existence, it appears to "carry" complex problems into another dimension for resolution. Answers are returned to our own. You might be surprised to learn that these matters have been presented in respectable journals for several years already.

This includes documenting the hypothesis that quantum entanglement occurs in the processing chip itself.[18] (Teleportation will be discussed by Josh in the final chapter of this volume).

It is my opinion that this computing method can help CERN accomplish its cosmic goals. Going one step further, it may be the tool employed to implement the "Mark of the Beast". All who accept this mark may be monitored *if not controlled* by this computer. For already at this moment, many assert it possesses a true artificial intelligence (AI). Recall, it was at CERN in 1989 British scientist Tim Berners-Lee is credited with having invented the World Wide Web. Originally it was intended for the sharing data generated at CERN (arising from its experiments and particle collisions) with researchers around the world. Today, this artificially intelligent AQC connects through the main hub at U.C. Berkeley, *to over 100,000 servers residing at CERN.* Through such networks as the ESnet5 (Energy Sciences Network 5) and the Helix Nebula Science Cloud, *the entire globe is now interconnected.*

"Linked in" are the massive data storage facilities of (so-called) *intelligence agencies of each major sovereign nation.* Thus, due to the interconnections of this computer through its network across borders,

---

[17] While controversial, many major companies doing research are looking carefully at such computers. Google discusses a recent finding in which the computing of the adiabatic quantum computer they have purchased, has successfully completed a comparative test of this type of processing with the classical processor. In this test, the **AQC was 100 million times faster than a classical computer chip.** See https://research.googleblog.com/2015/12/when-can-quantum-annealing-win.html.

[18] See the article from *Forbes* in 2013, written by Alex Knapp, retrieved from http://www.forbes.com/sites/alexknapp/2013/08/17/the-first-quantum-teleportation-in-a-computer-chip/#4e1b6a607fd9.

true sovereignty no longer exists. Instead, the one-world-order exists... *today.* What awaits us constitutes more than the opening of a portal and perhaps even *entities or at least their DNA transported adiabatically* into our realm, but the very revealing of the Anti-Christ. For two events will soon simultaneously occur as the crescendo of the agenda already in play today.

"POWER-UPS" OF MUCH GREATER MAGNITUDES

The power of these networked computers not only resides within the servers themselves, but of course the LHC too. Consequently, it's worthwhile to describe the actual mechanisms and consider their impact. The employed physics achieving the opening of this gateway is indeed entering the realm warned of by Dr. Hawking, and, I believe forbidden by God Himself.

This author announced Friday, October 23, 2015 on the Kev Baker Show, *Truth Frequency Radio Network*, that the LHC was designed and capable of achieving a power level of **1 PeV** (Peta electron Volts) to be obtained from the collisions of *heavy ions of lead particles.*

Then on November 26th, 2016, CERN made this announcement:

> Earlier this year, the LHC broke its own previous record by colliding protons at 13 TeV in the centre-of-mass, thanks to even higher-field magnets and an even larger ring. Its previous 7-8 TeV, before the improvements of the first long shutdown, was enough to create the Higgs boson.
>
> This week, just under three decades since the Tevatron reached 1 TeV, the LHC resumed its program of colliding so-called "heavy ions". More precisely, these are the nuclei of lead atoms. Since these nuclei contain the electric charges of 82 protons, the machine can accelerate them to 82 times the energy, reaching 1045 TeV in the collisions, breaking the symbolic barrier of a quadrillion electron-volts, or 1 PeV (Peta-electron volt).

Obviously, I woke up a few people by predicting what would happen just before CERN came clean and admitted huge new power levels.

For the reader's consideration, I have assembled the following information from both this author's and my friend, Kevin (Kev) Baker's websites. It may seem a bit technical, but a gist of this is all you need. *(See boxed material below.)*

So what do all of these numbers mean to us? Basically this: the actual MPL (maximum power level) that the LHC can achieve has been hidden from the public. We were kept in the dark until AFTER they completed their objective. All of the publicity beforehand had been surrounding having achieved **13 TeV**. No mention of **1.15 PeV**. It was not stated as an objective before being announced. It would seem that they are aware people like Anthony Patch are keeping an eye on them.

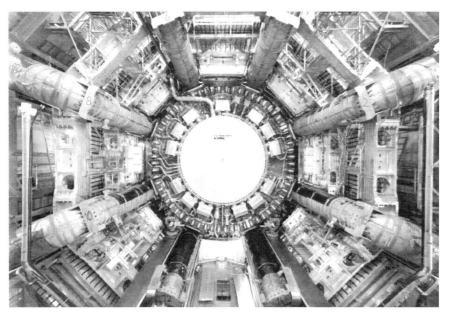

Figure 36 – THE ATLAS DETECTOR SEARCHES FOR OTHER DIMENSIONS

Recall: previously, they were colliding protons. On 26 November 2015 the type of collisions was dramatically changed, which generated much more powerful collisions, creating almost 1,000 times greater voltage than before by using ions of lead (Pb). After they had done this and lived to tell about it (and our planet was still spinning), they announced what they had achieved. Their announcement methodology doesn't give us much confidence that the public will be "kept in the loop", now does it? Forgiveness is always easier to receive than permission, so no wonder they followed this tact. Perhaps we need to remind them why kids are not to play with matches – they can get burned.

> The Maximum Power Level (MPL) for the LHC will be reached between November 20 – December 13, 2015: 1.15 PeV. As noted this exceeded the previous MPL of 13TeV. So how was this 1.15 PeV MPL achieved? As stated in the press release, it occurs through colliding lead (Pb) ions @ 1,148 TeV (1.15 PeV).
>
> Note: As stated at the outset of this chapter, the highest MPL previously announced was only 13 TeV utilizing protons (instead of ions).
>
> *To put in context:* Tera Electronvolt (TeV) is 10 followed by 12 zeroes or 10,000,000,000,000 (10^{12}). A Peta-electron (PeV) is 10 followed by 15 zeroes – 10,000,000,000,000,000 (10^{15}). This is known as the Center of Mass Energy (CME), derived from colliding lead particles. That is how the 1.14 PeV is to be understood. [19]
>
> The LHC increased its power up from 13 TeV to 1,148 TeV. Or to use all the zeroes to reveal the math to the common man or woman: 13,000,000,000,000 Electron Volts = the 13 TeV
>
> 1,148,000,000,000,000 Electron Volts = the 1,148 TeV = 1.15 PeV

CERN & THE FICTIONAL STARGATE: ARE THEY ALIKE?

"Portals to other dimensions… to make contact with other entities…" It brings to mind the movie and television show, *Stargate*.

Likewise, it sounds awfully reminiscent of the mission of the starship *Enterprise* with Captain Kirk at its helm:

> *These are the voyages of the starship Enterprise. Its five-year mission: to explore strange new worlds, to seek out new life and new civilizations, to boldly go where no man has gone before.*

---

[19] "The terms 'center of mass' and 'center of gravity' are used synonymously in a uniform gravity field to represent the unique point in an object or system which can be used to describe the system's response to external forces and torques. The **concept of the center of mass** is that of an average of the masses factored by their distances from a reference point. In one plane, that is **like the balancing of a seesaw about a pivot point** with respect to the torques produced." (Emphasis added, See http://hyperphysics.phy-astr.gsu.edu/hbase/cm.html)

As I've pointed out, leading physicists have expressed concern that this search for other dimensions could have disastrous consequences for us all. The warning has gone so far as to conjecture that vacuum bubbles could envelop not only our Earth, but perhaps the entire Universe. Citing from an NPR post on 19 February 2013,

> Back in 1982, physicists Michael Turner and Frank Wilczek wrote in *Nature* that "without warning, a bubble of true vacuum could nucleate somewhere in the universe and move outwards at the speed of light, and before we realized what swept by us our protons would decay away."[20]

As we discussed earlier, the production of *strangelets* – the most powerful explosive substance in the known universe – should sound the alarm bells too. The fact that the LHC will produce such a massive increase in power of 1.15 PeV, rather than the purported maximum of 14+ TeV., for 1,145 Trillion vs 14 Trillion is a big jump! To state it more explicitly:

- **PROTONS HAVE A COLLISION ENERGY = 14TeV**
- **LEAD ION COLLISION ENERGY = 1.15PeV**

The beam dynamics and the performance limits with lead ions are quite different from those of protons. This is due to the copious nuclear electromagnetic interaction in peripheral collisions of lead ions. The physics of lead ion beams is qualitatively and quantitatively different from that of Protons.

Plus, these lead ions are colliding *head-on*, whereas Protons do so at an angle. This contributes to the marked increase in the CME (Center of Mass-Energy level). To put it simply, that is why we see the jump from **TeV** to **PeV**. This increase in power, having been withheld from the public at large, calls into question their motives. "What will the result be? Why hide what you're doing?" Repeatedly, the spokespeople for CERN admit they have no idea what to expect once the LHC

---

[20] Peralta, Eyder, "If Higgs Boson Calculations Are Right, A Catastrophic 'Bubble' Could End Universe", *The Two-Way*, NPR, February 19, 2013. The article leaves the impression that the timing of the destructive bubble is unpredictable, that "another universe" could just pop-by and swallow us up. We'd never see what hit us since it would travel at the speed of light and it happen simultaneously with its appearing.

achieves maximum power levels which apparently are subject to magnitudes of change without warning. Really, no idea?

Again, just to emphasize the point here... we are talking about the difference between...TRILLIONS of electron volts (one million million) vs QUADRILLIONS (one thousand million million). That isn't subtle.

Imagine this scenario:

> Circulation within the 27-kilometer-long main ring of the LHC, not loosely configured subatomic particles... but now what essentially are two SOLID BEAMS OF LEAD, each rotating in opposite directions at 99.9999999991% the speed of light.

And, unlike what previously occurred with protons crossing each other's paths at an angle, heavy lead ions, the "solid beams", will collide head-on... Thus combining all of their kinetic energies at one point. It brings home what the Ghost Busters were talking about, when they shouted, "Don't cross the streams!" from their PROTON PACK particle accelerators as they were battling with Gozar![21]

With the above in mind, we finally come to the answer as to how this *power will be applied physically*. In short, and this is controversial, I believe *it will be focused upon a single point* (but wait a minute until I tell you where I think the target is). Situated 100 meters below grade and within the 27-kilometer Main Ring of the Large Hadron Collider are four principle particle detectors. These are: **ALICE, ATLAS, CMS** and **LHCb**. Here are descriptions of each:

- ALICE – A Large Ion Collider Experiment, producing quark-gluon plasma (*srangelets*) by colliding lead nuclei.
- ATLAS – A Toroidal LHC Apparatus, looking for particles or effects beyond Physics' so-called Standard Model.
- CMS – Compact Muon Solenoid, with the same purpose as ATLAS.

---

[21] "There's something very important I forgot to tell you! Don't cross the streams... It would be bad... Try to imagine all life as you know it stopping instantaneously and every molecule in your body exploding at the speed of light." — Egon Spengler (Harold Ramis) on the danger of crossing proton streams to contain negatively charged entities we call **ghosts**.

- LHCb – LHC beauty experiment, measuring certain B-hadron qualities such as asymmetries and CP violations.

It is within these principle detectors the *beams are crossed*. Each beam of particles circulates in opposing directions. The purpose is to collide particles upon their crossing, or in the case of heavy ions of lead, the actual head-to-head trajectory of particles moving *subluminal* – speeds just short the speed of light. Faster-than-light speeds have yet to be achieved (just a reminder in passing: don't forget the actions of *quantum entanglement* are much faster than light speed, virtually instantaneous even across vast reaches of outer space).

Most of the public announcements emanating from CERN have focused upon the collisions of protons. As stated at the outset of this chapter, during 2015, proton-to-proton collisions resulting in a record power of 13 TeV (Terra electron Volts) were produced by the crossing of particle streams within the detectors. Then in October through December of 2015, heavier ions of lead collided *not* at angles-of-crossing, (*indirectly* one might say). Instead, they were **direct**, i.e., *head-on collisions*. Thus, they realized much "higher-than-advertised" (more sarcastically, ever so "innocently" announced) levels of power in the **Peta** (that is, quadrillion) range – almost one thousand times more powerful.

Why is this significant? Remember that surpassing the threshold of 10 TeV produces *strangelets*, the most powerful explosive in the known Universe. Obviously, 1,145 trillion (Tera) far exceeds that of 10 trillion (Tera) electron volts by three orders of magnitude. To say the least, the priests at CERN achieved a quantum leap in the quantities of *strangelets* that "collided" into existence. And recall also that we said strangelets cannot be contained, not by the ALICE detector, or any other "holder" we might wish to try. Strangelets, like the Brothers Grimm's *Rumpelstiltskin* **fall right through the floor.** (When the princess called him by name, Mr. R., in has anger, stomped his right foot on the floor so hard he drove himself into a chasm never to be seen again!) In this case of the floor at the LHC, the strangelets go down deeper, *straight to the core of the planet*. Whereupon they begin attracting ordinary (unstable) matter to themselves, which – theoretically

– will one day convert the Earth into a neutron star. As I said at the outset, the time lapse before that happens may be as short as a decade and not likely any longer than a century. Happily, the Bible doesn't prophesy that our planet will collapse upon itself. However, those who don't believe in the Bible have a lot more to worry about, including strangelets and their affect upon our planet's longevity.

BEAM US DOWN SCOTTY!

Let's return to the notion of portals to elaborate a bit more on what it entails. Often, the question or statement arises regarding accessing other dimensions, or other "worlds" by spiritual means. Ancient peoples purportedly created portals. This seems to be what the Bible was saying. Therefore, if we believe the Bible (and our interpretation is correct), we know it was done before. Yet, we must ask, "Why again? Why now? Why would humanity require *the* most powerful device ever built be devoted to this task? What could justify the cost and the efforts of some of the best minds, from a hundred nations, for the past 70 years?" It must be very important to all of humanity. And this time, humanity must attempt to keep any god or demigod from getting in the way. So then, "What must be done this time around? What has to be different?"

Efforts must yield an opening different in *kind and duration.* This time, as was envisioned by the builders of the Tower of Babel, a stable opening must be manufactured that reaches into the heavens (far into the heavens) and isn't so easily knocked down. To clarify, while we continue to talk in terms of quantum scale, the portal must be more robust than what has been achieved to date in Geneva. The spiritual power to open a portal – the power apparently mustered by Nimrod was destroyed (perhaps working with technologies shown to him by fallen angels in the most-evil ever "transfer of technology"). And it was destroyed along with the confusing of humanity's tongues (the multiplication of languages, making global collaboration much more difficult). Humanity sought "to make a name for itself" – which may have been what Nimrod was seeking. *"And they said, 'Go to, let us build us a city and a tower, whose top may reach unto heaven; and let us make us a name, lest we be scattered abroad upon the face of the whole earth.'"* (Genesis 11:4)

Back in the day, about 4,000 years ago, give or take a century, it was likely *all about incantations* attempting to produce specific frequencies of vibratory sounds; or perhaps it involved *the ritual sacrifice of thousands of humans*; or possibly the issue was combining efforts – all humankind achieving a "concentration of consciousness" as the new-agers might say; that is, being focused together as one (known in the occult as an *egregore* which just happens to literally mean "watchers").[22] Whatever the method, it was to be employed in opening a window into the forbidden realm. We have good reason to suspect that they sought after, once more, *the ability to become gods*. The memory of the Nephilim and the whisper of more fallen angels likely made humanity jealous for powers that could keep them from being dispersed and kept at bay. To make a name for themselves may have meant just that – being like the angels and the Nephilim. "Hey everyone, let's be like the Watchers!" Whatever we put our mind to, we can get it done.

Fast forward four millennia, the vast numbers of coincidences, symbols, and pattern of occult rituals suggests to me that CERN scientists seek to hold open "the hatch" long enough for vast quantities of entities or the encoding of their beings to pass through into our realm and to be re-materialized on the other side of the transmission. In the next volume of our study, we will dig into this detail more. But, do make note: This is where the digitizing of DNA gets factored into this process. What I am talking about is the "transporting" of information about these entities at their present location through the portal and into our planet, specifically into the Abyss which the Bible says exists within this Earth. It very well may be the *objective of this secret sect* we've mentioned before, the

---

[22] "*Egregore* derives from the Greek word *egrégoroi*, meaning "watchers," which also transliterates as 'Grigori.'" It is used in the *Septuagint* in the *Book of Lamentations*, the *Book of Jubilees* and the *Book of Enoch*. See http://humanityhealing.net/2011/05/real-meaning-of-egregore/. The fact that Watchers could be linked to humanity's combined consciousness opens other interesting lines of discussion. But not here.

Brotherhood of Saturn, that these entities, like Captain Kirk and Mr. Spock from *Star Trek*, will manifest through the portal.

Admittedly, I am conjecturing something that others have not. But I am connecting some rather large dots, points already proven in the previous paragraphs. I believe that the highest level minders of CERN, whoever they are, lurk deep in the darkest corridor there or somewhere nearby in Europe, and they hold preternatural beliefs that would astound the average person. They hold (and this seems outrageous) that beings exist on the planet Saturn, the "better sun of the Golden Age", and they are being invited to return to our planet, which these modern-day sun worshipers believe was their planet before it was ours! But wait, I'm getting ahead of the story.

To be sure, the portal itself will be of a quantum scale – smaller than any atom. However, one must bear in mind, digital signals do not require great size to function. "Bandwidth" is not correlated to "broad width". The duration, coupled with the speed of transmission, is what matters. And what might be the subject of transmission is open to debate. But here is one such oddity that about which readers might wish to learn linking Saturn to UFOs and the possibility that some lifeform has made its presence known on the ringed planet.

Beginning with a book published in 1986 by Norman Bergrun called, *The Ringmakers of Saturn,* the "fringe" sought to explain observations by astronomers of bright lights on the edges of the rings from the past two centuries. Over the past 40 years, scientists began studying paradoxical photographs taken by the Voyager I and II that showed there appeared to be saucer-shaped spacecrafts of some sort, that were circumnavigating amongst Saturn's rings. Bergrun proposes that there are electromagnetic vehicles (EMVs) creating the rings of Saturn. Others in the crossover region between Ufology and the occult, speculate that intelligent life controls and cultivates the requisite substances for life on Saturn and interestingly, around our Sun. While we don't have more space here to elaborate on the many assertions of those in the so-called "woo-woo" world creating YouTube videos on the subject,[23] we can point out that it seems far more than coincidence that filmmaker Stanley Kubrick, obsessed with occult topics and encoding information about the secrets of the elites in almost all of his films,[24] implants a Saturnalia mystery in his film *2001: A Space Odyssey*. Some say that he feared his disclosure about the elite and their plans for humanity might get him killed. Why, for instance, did Hollywood make Kubrick change the target planet for the spacecraft operated by the Hal 9000 computer from Saturn to Jupiter? (Arthur C. Clarke's novel, *2001*, had *Saturn's moon Iapetus as the objective for the mission* – Saturn being far more laden with mythology than Jupiter). And what might be the real meaning behind the black monolith? The linkage to "the black cube" of Saturn? Many sources attest to the fact that the *black cube* has been being a symbol of Saturn down through the ages.[25] Once you begin to see the black cube linked to religions around the world, you begin to wonder if lying behind all the veneration of the black cube is the cult of Saturn and the secret society that celebrates Saturnalia. The material on this occult topic is legend. Here I would just say, "The plot thickens". We will no doubt delve into this matter more in our next volume. I promise to mention a few more points as we go along here, hopefully without getting us too sidetracked.

---

[23] *The Skeptics Dictionary* defines *woo-woo* as, "Woo-woo (or just plain *woo*) refers to ideas considered irrational or based on extremely flimsy evidence or that appeal to mysterious occult forces or powers. Here's a dictionary definition of woo-woo: "*Adj. concerned with emotions, mysticism, or spiritualism; other than rational or scientific; mysterious; new agey. Also n., a person who has mystical or new age beliefs."* When used by skeptics, woo-woo is a derogatory and dismissive term used to refer to beliefs one considers nonsense or to a person who holds such beliefs." Retrieved from http://skepdic.com/woowoo.html.

[24] Conspiracy-oriented filmmaker Jay Weidner made two films on this subject, *Kubrick's Odyssey 1 and 2*. The first film's subtitle: *Secrets Hidden in the Films of Stanley Kubrick.*

[25] "The lineage of the Illuminati traces back to mystery schools of the east. The cube represents earth in Pythagorean, Indian (Indian deities standing on cubes), Egyptian (Pharaohs sitting on cube thrones), and Platonic traditions. It's even realized as a sacred symbol today by the Freemasons (believed to be Illuminati affiliated) when they stand on an oblong square (cube) during particular rituals to the Worshipful Master. The cube is the building block of all nature, and the five solids of the ancients labeled as the "Pythagorean solids" include a tetrahedron (pyramid), cube, octahedron, icosahedron, and dodecahedron." See http://illuminatiwatcher.com/decoding-illuminati-symbolism-saturn-black-cube/.

We should not get hung up on whether such entities that might be the targets of particle beams from CERN are "spiritual" or "physical". Nor should we attempt to fathom the mathematics involved or the processes in any depth to what may be the goal for connecting CERN to Saturn. Furthermore, we do not need to stipulate that its technology will in fact work as advertised. We should recognize, however, that the geniuses I conjecture are attempting this *manifesting miracle* seem earnest to make it so. It is their passion – it is their religion – it is their raison d'être.

Wave-particle duality provides an analogy here. Every elementary particle, the physical may be partly described in terms not only of particles, but also of waves. Quantum-scale objects cannot be fully expressed using classical-model concepts of our Universe. As Einstein wrote: "It seems as though we must use sometimes the one theory and sometimes the other, while at times we may use either. We are faced with a new kind of difficulty. We have two contradictory pictures of reality; separately neither of them fully explains the phenomena of light, but together they do."[26] Two pictures of reality that contradict and yet they are both proven to be true. Illogical perhaps, but true. Even scientists must embrace what Theologians call *antinomy*.[27]

The power of *quadnary computing* by the adiabatic processor possesses the means to accomplish what traditional processors cannot. Mathematical riddles that take our most powerful classic computer 30 minutes to solve, can be unraveled in less than 3 seconds. In fact, the proper order of magnitude would be more like .00003 seconds. In certain combinatorial computations, the Adiabatic Quantum Computer has proven *its processing power may be 10 million times faster than a binary processor!*

---

[26] Harrison, David (2002*).* "Complementarity and the Copenhagen Interpretation of Quantum Mechanics". *UPSCALE.* Dept. of Physics, U. of Toronto.

[27] Antinomy is the apparent mutual contradiction of two laws that are both true. Free will and predestination comprise a Theological antinomy. Immanuel Kant references four key antinomies in his philosophy. Beyond our scope, but that discussion is intriguing. See https://en.wikipedia.org/wiki/Antinomy to consider Kant's antinomies that crossover between the realms of Science and Theology.

FROM THE RINGS OF CERN TO THE RINGS OF SATURN

Recall from earlier, the formation by the main ring of the LHC of a doughnut-shaped toroidal field. This mimics those surrounding Earth, providing us shielding from gamma rays, X-rays and accelerated particles streaming in from the sun and Cosmos at large. Earth's magnetosphere and magnetopause provide this shield.

As well, we must not forget the AWAKE (Advanced Wakefield Experiment) linear particle accelerator housed within CERN. It will produce a straight-line stream of accelerated heavy ions of lead, at a level 1,000 times the power to date achieved by the LHC alone.

This powerful stream will pass through a torsion field formed by the toroidal field of the LHC. Just as the near-the-speed-of-light contra-rotating particle beams are squeezed down by the focusing actions of magnetic collimators housed within the three-story tall detectors, likewise the linear beam produced by the AWAKE device will be aimed by the niobium-titanium superconducting magnets of the LHC. Magnetic lines of force, at 8.36 Tesla, or 100,000 times that of Earth's magnetic field, form up torsion fields through which a linear stream of accelerated particles approaching the speed of light will be aimed at a specific target in the outer reaches of the Solar System.

So focused is this beam, that it will be dialed in to strike at a locale of the sub-atomic quantum scale millions of miles from Earth, far beyond our magnetosphere, which in fact, must be reduced in strength to allow the beam to pass through ("Shield's down", Mr. Sulu).[28]

AWAKE's linear stream of accelerated particles combined with the power of the LHC will be aimed with extreme accuracy. A gap between quantum particles will be created. How long can the priesthood of scientists maintain such an opening? This is where the Adiabatic Quantum Computer achieves its awesome potential. Already operating at the quantum level in space as well as a quantum scale of time, the portal will be opened for what a human would swear is but a few moments, while a

---

[28] Separately, this beam might be used in the "Star Wars of all Star Wars", as a particle beam weapon focused on a returning Messiah to kill Him and His army of "mighty ones".

sensitive detector would regard as a constant, never-ending aeon (lasting a lifetime in quantum terms). The beam emitted through the ring of CERN accompanied by the AWAKE linear accelerator will draw a bead on Saturn, past its rings to a very specific point on a planet far, far away.

*Now, here is where I go into **full** speculation mode.* This is a possible scenario and the reader should not take my speculation as fact, for at this time I don't have the evidence to do more than connect the dots with faint, dashed lines instead of heavy, coarse lines. And consider this scenario as archetypal of the kind of thing that the masters of the machine might be trying to do. Here goes:

Figure 37 – THE HEXAGONAL FORMATION ATOP SATURN: A SIX-SIDED CLOUD PATTERN - A CLUE TO A SATURNALIAN OUTCOME FOR US?

Humanity's most expensive machine – perhaps its greatest mechanical accomplishment – will target the southern pole of Saturn. By connecting to and opening a quantum gap at this very specific point in our solar system, it gains access (remember I'm speaking hypothetically), to a very particular container awaiting a wakeup call on the colossal planet. Like a key slid gently into lock, the beam from CERN will connect with a *magnetic black cube* (perhaps "magical" would be the better word here), within which a worshipping cult claims are imprisoned the Titans of Old – the Masters of the Universe. Like Zod and his fiendish accomplices who were banished from Krypton and sent into the "Phantom Zone" entrapped in a marvelous "container" to be held for tens of thousands of years, (recalling the mythology of *Superman*), the practitioners and worshippers of "the Black Sun" – members of one of the most secret of secret societies – will then exult in accomplishing what has seemed for ages to be beyond humanity's grasp. They believe this solemn black cube on the Golden Sun of Saturn imprisons those whose powers will provide humanity immortality – the power to make men gods! The black cube is not just a symbol of Saturn or its orbit as Kepler projected, there is a black cube within the planet of Saturn.

At the northern pole of Saturn is located what many observers would consider to be a naturally occurring formation of contra-rotating clouds of energy and particles undergoing acceleration. The origin of this six-sided formation is debated. This author contends it to be purposeful in both its design and presentation. It appears as a larger-than-Earth version of the Large Hadron Collider. In fact, the LHC functions in mimicry of it. Like the LHC, the northern pole of Saturn is generating Synchrotron-like energies of gamma and X-rays. Additionally, it shows electromagnetic lines of force. These energies may extend all the way to the planet's core to where, in my "vision", lies the magnetic black cube. Extending beyond and wrapped around the planet, these lines of rotating forces terminate at the southern pole manifesting as a spiral, not unlike that which was observed in the night sky on December 9, 2009, over Norway. Looking back in light of this plan, we might wonder, "Was that spiral a sign in the heavens, a portent of this future event?"

This is where the prodigious priests aim that 27-kilometer-long weapon in hopes to spring open the black cube, freeing those with whom they've been communing "on the other side" (and beyond the orbit of Jupiter). For they believe by so doing, their "ancestors from the stars", their "benevolent entities" will return along pathways of electrical Birkeland currents to Earth. Like ET phoning home, the beam will serve as more than a homing beacon – it will provide the means by which the ancients will return. The linear accelerator AWAKE, when combined with the torsion field of the LHC will produce the characteristic twisted, helical double-strands of charged particles (observed by the Hubble space telescope connecting concentrations of electrically charged plasma throughout the Cosmos – recall our earlier discussion).

Continuing with the conjecture: Birkeland currents serve to transmit the digitized DNA of demonic Titans of old directly to and then below Earth's surface. Shockingly, the entry point is at CERN, where perhaps the shaft to the Abyss as described in the Bible resides.

Figure 38 - BIRKELAND CURRENTS IN THE COSMOS

Therefore, it appears this was the purpose that the LHC was located on the ancient site of the Roman temple to their god Apollo, Apollyon, Abaddon, the King of the Abyss, also known as "The Scorpion King". Revelation 9:11 says. *"And they had a king over them, which is the angel of the bottomless pit, whose name in the Hebrew tongue is Abaddon, but in the Greek tongue hath his name Apollyon."* The name means destroyer. Revelation 9:1 says. *"And the fifth angel sounded, and I saw a star fall from heaven unto the earth: and to him was given the key to the bottomless pit."*

It is another non-coincidence that Birkeland currents form helical strands as does DNA. In combination with the AWAKE particle accelerator, the *LHC would then become the key to the Abyss*, unlocking its shaft. Perhaps by transmitting the DNA of demons, through its encoding and decoding by the Adiabatic Computer, these prisoners from eons long since passed will awake. This sequence of information will be teleported to beneath their ancient temple to the Abyss atop where CERN sits.

It sounds like an H.P. Lovecraft novella, but the end goal of the elites might look a lot like this scenario. Consider… How remarkable. The advances in humanity's science have coalesced into this. The secret societies down through the ages have succeeded in delivering to our highly technological age and to some of its greatest minds, the mind-boggling know-how **to attempt a cosmic jail break.**

Genesis 6:5-6, *"And God saw that the wickedness of man was great in the earth, and that every imagination of the thoughts of his heart was only evil continually. And it repented the LORD that he had made man on the earth, and it grieved him at his heart."*

CONCLUSION

Obviously this secret cult scenario begs the question, "If this truly comprises the great plan behind CERN, can anything be done to expose it? Can the scheme of the priests be thwarted? What can we do?"

Perhaps it is far too late to rescue the world from the greatest conspiracy in our Cosmos. The challenge for us: *The scheme is so far-fetched that few will believe it.* Hollywood couldn't make this up, but they have tried.[29] Just as the massive scale of the Holocaust foisted by the Nazis on the Jews (and other victims) made it nearly impossible to believe, so the colossal, indeed cosmic aspects of this plan would be met by almost all with incredulity. But realize this: just because it is unbelievable, that doesn't mean it, or something like it, isn't still true. Certainly, we know there are those who do believe in occult power (like the Brotherhood of Saturn) will seek to make it manifest in the real world.

Nevertheless, it is my hope that to the extent we can have some effect, we must attempt to cast light upon humanity's dark conspiracies, fully

---

[29] For a particularly entertaining romp through the subliminal encoding of the black cube in Hollywood, visit https://hollywoodsubliminals.wordpress.com/black-cube/.

aware that men loved darkness instead of the light. *"And this is the condemnation, that light is come into the world, and men loved darkness rather than light, because their deeds were evil."* (John 3:19)

This author finds solace in the following, Ephesians 6:10-18...

> *[10] Finally, my brethren, be strong in the LORD, and in the power of his might.*
>
> *[11] Put on the whole armour of God, that ye may be able to stand against the wiles of the devil.*
>
> *[12] For we wrestle not against flesh and blood, but against principalities, against powers, against the rulers of the darkness of this world, against spiritual wickedness in high places.*
>
> *[13] Wherefore take unto you the whole armour of God, that ye may be able to withstand in the evil day, and having done all, to stand.*
>
> *[14] Stand therefore, having your loins girt about with truth, and having on the breastplate of righteousness;*
>
> *[15] And your feet shod with the preparation of the gospel of peace;*
>
> *[16] Above all, taking the shield of faith, wherewith ye shall be able to quench all the fiery darts of the wicked.*
>
> *[17] And take the helmet of salvation, and the sword of the Spirit, which is the word of God:*
>
> *[18] Praying always with all prayer and supplication in the Spirit, and watching thereunto with all perseverance and supplication for all saints.*

What has been presented here will ultimately work out according to God's plan. Exactly how the mythos of Saturn ties in remains to be seen. But soon, Satan and his minions will be cast into the lake of fire.

> *[10] And the devil that deceived them was cast into the lake of fire and brimstone, where the beast and the false prophet are, and shall be tormented day and night for ever and ever.*
>
> *[11] And I saw a great white throne, and him that sat on it, from whose face the earth and the heaven fled away; and there was found no place for them.*
>
> *[12] And I saw the dead, small and great, stand before God; and the books were opened: and another book was opened, which is the book of life: and the dead were judged out of those things which were written in the books, according to their works.*

¹³ *And the sea gave up the dead which were in it; and death and hell delivered up the dead which were in them: and they were judged every man according to their works.*

¹⁴ *And death and hell were cast into the lake of fire. This is the second death.*

¹⁵ *And whosoever was not found written in the book of life was cast into the lake of fire. (Revelation 20: 10-15)*

WAS EINSTEIN WRONG?
THE EMERGING COSMOLOGY—PHYSICS, ALCHEMY, AND THE NEW REALITY[1]

S. Douglas Woodward

> There's nothing religious in any of these matters,
> In the superstitions or in the prophecies
> Or in anything that people call the occult—
> Above all there is a way to look at nature
> And a way to interpret nature
> Which is completely legitimate.
>
> *Guillaume Apollinaire (1890-1918)*

> The Humbug[2] is not the man who dives into mystery
> but the one who refuses to come out of it.
>
> *G.K. Chesterton (1874-1936)*

SCIENTIFIC REGARD FOR THE SUPERNATURAL

CONVENTIONAL THINKING ASSUMES "EVERY CAUSE HAS AN EFFECT" WHICH SPECIFIES THAT WHATEVER HAPPENS CAN BE EXPLAINED WITHOUT RECOURSE TO THE SUPERNATURAL OR THE miraculous. The average person on the street today, supposes most scientists continue to subscribe to this enlightenment way of thinking. However, old conceptions of reality which dismissed esoteric phenomena (a

---

[1] This chapter was originally published in *Power Quest, Book One: America's Obsession with the Paranormal*. It has bene revised and updated for this book.

[2] Something "silly or makes no sense; something... meant to deceive or cheat people." (Encarta Dictionary)

point of view so dominant among the intelligentsia for more than 300 years) are now being questioned by serious thinkers.

This is so despite skeptics like Richard Dawkins of the so-called *Brights Movement* who represent a diminishing number of noisy atheistic naturalists, in select scientific and secular circles, which reject the God of the Bible (or even the deistic/gnostic *Supreme Being* of Freemasonry). Nevertheless, when one scans the panorama of various philosophies in play today, there are in fact far fewer skeptics. When it comes to the willingness of most philosophers and intellectuals to, shall we say "expect the unexpected"; many smart people find reasons to believe in the mystical. Mystery, if not mysticism has found a place in the twenty-first century among many scientific scholars, especially those entrenched in exotic technology. It turns out that *the fresh brute facts of physics encourage a scientific form of spiritualism*—however paradoxical such a notion may be.

Figure 39 - THE MORNING OF THE MAGICIANS

Louis Pauwels and Jacques Bergier note in their classic *The Morning of the Magicians*:

> Physics unveils a world which operates on several levels at the same time and has many doors opening on to infinity. The exact sciences [physics or astronomy] border on the fantastic. The humane sciences such as psychology or sociology] are still hedged about with positivist[3] superstitions. The notion... of evolution dominates scientific thinking.

---

[3] *Positivism* is the "theory that knowledge can be acquired only through direct observation and experimentation, and not through metaphysics or theology" according to the *Encarta Dictionary*.

Although their view was penned in 1960, their observation arguably has been proven correct. Not only is their 50-year-old opinion an accurate assessment of the various sciences of their day, but a valid prediction of the contemporary situation. And yet, we should realize there have always been voices challenging the skeptics of naturalism[4].

Under-publicized if you will, are numerous respected declarations throughout the past 300 years that present "the minority report" – questioning whether such skeptical certitude about nature and normality is justified. Goethe,[5] a poetic voice during the Age of Enlightenment, offered a strongly worded, contrarian opinion: "We walk in mysteries. We are surrounded by an atmosphere about which we still know nothing at all. We do not know what stirs in it and how it is connected with our intelligence. This much is certain, under particular conditions the antennae of our souls are able to reach out beyond their physical limitations."[6]

Today, this advocacy for the abnormal grows even stronger. The esoteric steadily encroaches upon the fortress of skepticism built over the past three centuries during *naturalism's* hegemony. Not since *before* the Enlightenment with the *Neo-Platonists of the Renaissance* has the supernatural made such broad headway amongst intellectuals in Western society. These Neo-Platonists of the Renaissance period were particularly important in seeking to develop a cosmology that blended mysticism and magic into Christianity, creating a supernatural cosmology similar to Gnosticism in ancient times and Shamanism today.

Pico del la Marandola was noted for his attempt to develop a Christian form of Cabbala. *Giordano Bruno* was excommunicated by Catholic and Protestant churches alike for his occult beliefs. *Heinrich Cornelius*

---

[4] *Naturalism* is the view that all effects are the result of natural causes—the paranormal has no place in reality whatsoever.

[5] Johann Wolfgang von Goethe (1749–1832), mentioned earlier, was the genius of German literature and author of *Faust*.

[6] As quoted by Daniel Pinchbeck, *Breaking Open the Head: A Psychedelic Journey into the Heart of Contemporary Shamanism* (New York, NY: Broadway Books, 2002), 112.

Agrippa wrote *De Occulta Philosophia Libri Tres.*[7] A brief synopsis in *Wikipedia* states,

> Agrippa argued for a synthetic vision of magic whereby the natural world combined with the celestial and the divine through Neo-platonic participation, such that ordinarily licit[8] natural magic was in fact validated by a kind of demonic magic sourced ultimately from God. By this means Agrippa proposed a magic that could resolve all epistemological problems raised by skepticism in a total validation of Christian faith.[9]

Bruno, Marandola, and Agrippa were thoroughly "renaissance men" living in the sixteenth and seventeenth centuries. No doubt, they were illuminated men that would have remained committed to a mystical view of reality even if, hypothetically speaking, they would have encountered the enlightened skeptics of the eighteenth and nineteenth centuries who found no reason whatsoever to believe in such mumbo jumbo.[10] So while the majority of intellectuals following Bruno, Marandola, and Agrippa dismiss any sense of the supernatural (thereby eliminating the chance for the miracles of Christianity), they opted to throw out the baby with the bath water. Except for the contrarians noted, doubt and skepticism would hold serve from thenceforth to our own day.

Likewise, mainstream Christian theology would evolve in Europe into a *faith without miracles* except in selected evangelical pockets. America's mainline denominations too, in the twentieth century

---

[7] *Three Books Concerning Occult Philosophy*, Book 1 printed Paris, 1531; Books 1-3 in Cologne, 1533.

[8] Licit in this context conveys something allowed by God's law.

[9] See *http://en.wikipedia.org/wiki/Heinrich_Cornelius_Agrippa*.

[10] The memory of these Renaissance men should not be metaphorically burned at the stake for divination even if all that is at risk is their reputation. Why? They were sincerely seeking a place for faith and miracles. They hoped by citing occult examples and unifying the occult worldview with the Christian, they could make "sense" of the biblical cosmology to which they were committed. It would offer "proof" for the supernatural, and thus, the Bible. We should be sympathetic to their goal, although their thinking and approach falls outside the bounds of biblical orthodoxy. It is instructive that Jesus received vocal validation from demons that He was the Son of God—despite the cloak of human flesh. Jesus refused to accept their acknowledgement. He commanded they be quiet. The miracles of the occult are not to be sought as evidence that the Bible is true, although they are evidence enough of the supernatural.

would disparage the supernatural and depart from the true faith. Consequently, with the about face of physics later in the twentieth century and now in the twenty-first – seemingly eager to embrace a much more mystical point of view – it is all the more astounding that the new found faith springs forth from the "exact sciences" and not from mainstream theology which remains hopelessly lost in demythologizing theological content. Mainstream Christian theology does little more in our day than stumble around in search of a reason to affirm some type of God-concept along with justifying the value of faith in Him – or more accurately – *it*. The author laments that just when the game is going down to the wire, mainstream Christianity is hopelessly feckless, unable to enter the fray as an "impact player". Having spent most of my Christian life in such churches (with good people, but bad theology), my frustration on this point is profound.

THE NEW REALITY

Pauwels and Bergier's primary purpose in writing *The Morning of the Magicians* was to build a case for "fantastic reality" – a cosmology explaining the universe which overcomes the limitations of the positivist approach; whereby it incorporates the new physics, modern mathematics, and opens the door to the realization that *the natural and supernatural may not be so distinct.* In short, *many aspects of this "super-nature" aren't supernatural at all.* There need be no recourse to supposed actions of other-worldly beings to explain the behavior of physical objects.

However, by itself, this perspective has *nothing* to do with whether or not God created the heavens and the earth; and yet, it has everything to do with our ability to fully comprehend how (from this author's vantage point) God [11] created (and sustains) His Universe. In other words, the issue is we simply don't understand physics and mathematics well enough to plumb the wisdom or methods of our Creator.[12]

---

[11] This God could be the God of the Bible, the 'Supreme Being' of Freemasonry, or a pantheistic notion. I am not arguing that God necessarily has to be the creator Yahweh although my personal belief is that He is.

[12] Of course, we who are evangelicals identify God as incarnated in Jesus Christ.

Figure 40 – NIKOLA TESLA READING ROGER BOSCOVITCH'S BOOK,
Theoria Philosophiae Naturalis

As part of their closing argument, Pauwels and Bergier tell a story of a Serbian genius, Roger Boscovitch (*Ruđer Josip Bošković*, 1711-1787) who was two centuries ahead of his time in these precise sciences. In relating his story, they make the point that the human mind is capable of far greater insightful leaps than we might suppose. Indeed, someone like Boscovitch was way too advanced for his peers. The challenge for Boscovitch was not so much that his peers doubted he could obtain genuine breakthroughs. At fault was that without societal support, such revolutionary insights (which contradicted contemporary conventional wisdom of his times) would not be acceptable to the science of that day and thus of necessity lie dormant (on the shelf, as it were) for centuries before others would arise smart enough to fathom such concepts. It turns out this is precisely the case with Roger Boscovitch and his cosmologic model of the universe. A brief recital here of his story will be helpful to our study.

By the time he was 29, Boscovitch was already a teacher of mathematics in Rome and a science advisor to the Pope. On June 26, 1760 at 49 he was elected a Fellow of the Royal Society in England, publishing on that occasion a poem on the visible features of the sun and moon that had his colleagues exclaiming, "This is Newton speaking through the mouth of Vergil [*Virgil*]." Only in the 1950s, at the behest of the Yugoslav Government [today's Serbia], had Boscovitch's *Theory of Natural Philosophy* (1758), began to be seriously reconsidered.

It seems that Boscovitch was in advance, not only of the science of his time, but of our own [surpassing Einstein's theories which apparently cannot be integrated with the Quantum mechanics theory of Heisenberg]. He [Boscovitch] proposed a unitary theory of the Universe, a single general and unique equation governing mechanics, physics, chemistry, biology, and even psychology. According to this theory, matter, time, and space are not infinitely divisible but composed of points or grains… We find in his works the quanta, the wave mechanics, and the atom formed of nucleons. The scientific historian, L.L. Whyte, assures us that Boscovitch was at least two centuries ahead of his time, and that we shall only be able to understand him when the junction between relativity and quantum physics has finally been effected. It is estimated that in 1987 [Pauwels and Bergier wrote this prediction in 1960], on the two hundredth anniversary of his birth, his work, will be appreciated at its true value.[13]

Pauwels and Bergier proclaim victory for their thesis through the efficacy of *mathematics* – proof positive their hope (to restore among us an appreciation of the "fantastic reality" that is nature) is becoming more compelling with each day that passes, as mathematics continues to achieve new breakthroughs demonstrating their view to be correct. They summarize this proclamation with these words:

> The language of modern mathematics is the only one, no doubt, that can give some account of certain results of analogical thinking. There exist in mathematical physics regions of the "Absolute Elsewhere" and of *"continus de mesure nulle,"* that is to *say measurements applied to Universes that are inconceivable and yet real.* We may wonder why it is that the poets have not yet turned to this science to catch an echo of the music of those spheres of *fantastic reality*—unless it be for fear of having to accept this evidence – that the magic art lives and flourishes outside their study walls.
>
> The mathematical language which is proof of the existence of a Universe beyond the grasp of a normal waking consciousness is the only one that is in a state of constant ferment and activity.[14] [Emphasis mine]

To Pauwels and Bergier, the proof isn't in the pudding but lies in the recipe (more accurately, the *equation*) that explains why the *Cosmos* behaves the way it does. As they say, mathematics continues to ferment relentlessly, eventually righting the wrongs in formerly mistaken methods of how we perceive reality.

THE STANDARD THEORY PERSEVERES

However, while new mathematical vistas may pave the way to overturn cherished beliefs about gravity, light, space, and time, as we study the writings of various physicists over the past century, we may

---

[13] Louis Pauwels and Jacques Bergier, *The Morning of the Magicians,* p. 351. This prediction may be slightly too optimistic—his theories however, are consistent with other non-standard physicists, as we will see in our study.

[14] Ibid., pp. 322-323.

soon conclude that what one scientist finds compelling another mistrusts. Despite the inconsistencies their two models (Einstein and Heisenberg), the *standard theory* nevertheless perseveres. However, even the standard theory is nowadays further away from consensus. For instance, we know that Nikola Tesla (1856-1943, see earlier photo), financed by J.P. Morgan, never agreed with Albert Einstein's model regarding the relativity of time and the possibility that space could be bent by mass. Tesla perceived the physical phenomenon that Einstein was describing to be caused by undiscovered properties of time and space unknown to Einstein. Other standard axioms of physics (the well-celebrated laws of *thermodynamics*), such as the *indestructibility of matter*, the law of *conservation of energy* (energy can change forms but there is no loss of energy); even these *laws* Tesla believed are subject to challenge – they are not immutable – they may even be outright wrong.

To be specific, what is now being considered is this: in our universe, the mass of an object can fluctuate depending upon its surroundings as intersected by the properties of space or time at that place and moment. Additionally, entropy (aka heat loss, increase in randomness) is presumed but not consistently observed countering what most scientists would predict. As we will see below, time itself yields an energy that may serve as a preservative of sorts to processes; this energy may overcome entropy forestalling what is known as thermal death. In short, our assumptions about reality (and therefore, Einsteinium physics) may have focused on the wrong aspect of space-time. Therefore, even the supremely sacred laws of thermodynamics may be erroneous.

This really shouldn't surprise us. The final word in physics has not been spoken. The possibility for one cosmological model to trump another has been famously seen numerous times. We know Einstein's model explained certain deficiencies in Newton's laws of gravity (commonly expressed as stating that time and space are not absolute and fixed, but *relative and subject to fluctuation*). Likewise, Heisenberg theorized that matter was composed of *quanta* (think of *particle* as a simple analogue) at its lowest level whose speed and location could not both be determined through measurement *without affecting the quanta* (reality appears like a wave or field instead of a particle). Known as *Heisenberg's Uncertainty Principle*, physics has implied ever since that science is limited in its ability to measure reality (at the sub-atomic level) because science couldn't predict with exact certainly the properties of the quanta. Furthermore, with the cast-iron certainty of *Planck's constant* (there is a limit to how small a particle can get in our universe), and the measurement of the total mass in the Universe (we can in fact accurately calculate the number of atoms in the Cosmos); we know that *the universe is not infinite on either end of the spectrum*. We live in a finite Cosmos. It may be growing, but it had a definite beginning and it does not extend to 'eternity'. In the final analysis, what was presumed to be true about nature in 1900 was passé by 1910. What was true in 1910 was passé by 1925. And what we've believed ever since that point in time may be ready for capsizing today as well.

Science has searched feverously (beginning with Einstein and ever since) for what is called a *unified field theory*. In simplest terms, the idea is that every attribute or property about space, time, matter, energy, (and the distinction of the various fundamental forces and particles) can be expressed and held to be true in every single point in the Cosmos (reflect on your high school geometry and picture a specific coordinate point defined *dimensionally*—the old "x-y axis" to be precise). The unified field theory seeks to explain both the usual and the unusual—everywhere, every time.

> [So far] there is no accepted *unified field theory*, and thus [it] remains an open line of research. The term was coined by Einstein, who attempted to unify the general *theory of relativity* with *electromagnetism*, hoping to recover an approximation for quantum theory. A *theory of everything* is closely related to unified field theory, but differs by not requiring the basis of nature to be *fields* [think of a magnetic field to get the idea], and also attempts to explain all physical constants of nature.[15] [Emphasis and comments mine]

---

[15] See http://en.wikipedia.org/wiki/Unified_theory. Further detail: "According to the current understanding of physics, forces are not transmitted directly between objects, but

It is fascinating that this 'new physics' is not really new. Its life in the shadows was partially because, in the case of Roger Boscovitch, *the new reality* didn't achieve the notoriety to become the *standard* theory. At the very least, Boscovitch lacked a publicist. (Or more precisely, someone like Tesla who grasped the point that Boscovitch was making). But in the case of the next physicist we will study ('in the dock' as it were), his obscurity was *intentional*—stemming from a top-secret status his theories held in the Soviet Union.

IS IT TIME FOR THE AETHER - AGAIN?

According to theologian, Joseph P. Farrell[16] in his book, *The Philosopher's Stone* (2006, subtitled, *Alchemy and the Secret Research for Exotic Matter*) the work of a particular Russian during the 1950s was a breakthrough in providing a predictive model *that is unified* not only explaining physics but also plumbing the depths of human consciousness too. This physicist was **Nikolai Kozyrev** (1908-1983).

Farrell points out that Nikolai Kozyrev's theories appear to coincide with Nikola Tesla's assertions (and we might add appear to support Roger Boscovitch's work as well—note the figure earlier where Tesla studies

---

instead are described by intermediary entities called fields. All four of the known fundamental forces are mediated by fields, which in the Standard Model of particle physics result from exchange of gauge bosons. Specifically, the four interactions to be unified are:

Strong interaction: the interaction responsible for holding quarks together to form neutrons and protons, and holding neutrons and protons together to form nuclei. The exchange particle that mediates this force is the gluon.

Electromagnetic interaction: the familiar interaction that acts on electrically charged particles. The photon is the exchange particle for this force.

Weak interaction: a repulsive short-range interaction responsible for some forms of radioactivity that acts on electrons, neutrinos, and quarks. It is governed by the W and Z bosons.

Gravitational interaction: a long-range attractive interaction that acts on *all* particles. The postulated exchange particle has been named the graviton.

Modern unified field theory attempts to bring these four interactions together into a single framework..."

[16] Farrell is a professor of Patristics (the study of the Church Fathers) at California Graduate School of Theology. He has authored books on the East-West Schism in Christianity and books of alternative history and science. He received a M.A. degree at Oral Roberts University and doctorate at Pembroke College, Oxford.

Boscovitch's book). The reason that Kozyrev is not a household name like Einstein stems from the Soviets who kept his studies classified *top secret*. Only in the past few years have his theories become public.

A spiritualist researcher, David Wilcock[17] is cited by Farrell as follows, "The awesome implications of his work and of all those who followed him, were almost entirely concealed by the former Soviet Union..." Farrell continues, "In other words, Kozyrev's work was so awesome and extraordinary in its implications, not only for the development of the foundation of theoretical physics, but also for its dangerous potential applications, that the Soviet leadership wisely classified it at the very highest levels."[18]

Figure 41- NIKOLAI KOZYREV

Then Farrell cites Wilcock twice more in the following paragraph to tie the theories of Kozyrev and Tesla together:

> Tesla, Wilcock notes, stated in 1891 that the physical medium "behaves as a fluid to solid bodies, and as a solid to light and heat." Moreover, with "sufficiently high voltage and frequency" the medium itself could be accessed. This was, as Wilcock correctly notes, Tesla's "hint that free energy and anti-gravity technologies were possible." [Of course, free energy didn't please Tesla's patron, J.P. Morgan, owner of Standard Oil.] It is Tesla's assertion that the medium has fluid-like properties

---

[17] David Wilcock is a researcher and author, with a Bachelor's in Psychology, with numerous books and movies to his credit. David believes that he may be the reincarnation of Edgar Cayce and has written material to support this (which this author is simply sharing here, without passing judgment). He has also made a number of public predictions including the JFK Jr. death (the night before on radio) and the 911 disaster. See his credentials at his web site: http://divineCosmos.com/component/content/158?task=view. I respect his point of view, but do not endorse it.

[18] Farrell, Joseph P. *The Philosophers Stone, Alchemy and the Secret Research for Exotic Matter*, Port Townsend, WA. Feral House, 2009, pp. 151-152. Quoting David Wilcock, "The Breakthroughs of Dr. Nikolai A. Kozyrev," *The Divine Cosmos*, www.divineCosmos.com, pp. 1-2.

that ties his work directly with the work and thought of Dr. Kozyrev. [Comments added]

What does Farrell mean by *medium*? The *medium* means that whatever *it* is, *it* is the most essential aspect of reality, the most basic building block, if you will. As such, this constitutes the key scientific (some would say, quasi-scientific) issue to grasp in this chapter. Therefore, I encourage the reader to pay careful attention!

Throughout history, scientists and alchemists alike – from alchemists of ancient times like *Trismegistus* (ca. 300-150 BC) to *Paracelsus* (1493-1541), to the German Franz Anton *Mesmer* (1734-1815, from which we derive the word *mesmerized*), to today's para-psychologists – have suggested that there is a discrete medium that extends throughout the universe which is actively involved in influencing mass and energy. It is often called, the *aether*. The very foundation of *the dark art of alchemy is based upon this notion.* (As discussed in a previous chapter, the Electric Universe endorses the notion of *aether* as a neutrino-based plasma.)

> Aether theories in early modern physics proposed the existence of a medium, the *aether* (also spelled *ether*, from the Greek word—αἰθήρ-*aether*—meaning "upper air" or "pure, fresh air"), a space-filling substance or field, thought to be necessary as a transmission medium for the propagation of electromagnetic waves. The assorted *aether theories* embody the various conceptions of this *medium* and *substance.* [19]

The notion of *the aether* has been hotly debated over the centuries as the reader would imagine. Most scientists believe that Einstein rejected it (assuming that it was not *elegant* enough to be true). Robert Youngson's 1998 article in *Scientific Blunders* asserted this point of view: "By 1930, younger physicists would smile in a supercilious fashion at any reference to the aether. All scientists now that, in the words of the American homespun philosopher: 'There ain't no such

---

[19] See http://en.wikipedia.org/wiki/Aether_theories.

critter.'"[20] However, David Wilcock challenges this traditional characterization of Einstein. By 1918, Einstein was swayed to believe in a *medium*. We read:

> [Any] part of space without matter and without electromagnetic fields seems to be completely empty... [But] according to the general theory of relativity, even space that is empty in this sense has physical properties. This... can be easily understood by speaking about an ether, whose state varies continuously from point to point. [Brackets in Wilcock's citation][21]

Despite what Youngson contends, other physicists are reintroducing it as it explains some phenomena that are now observable – for instance, *how particles can affect one another at rates millions of times faster than the speed of light!*

Kozyrev (and apparently Tesla) believed that "time" itself *was in fact this aether*. As Farrell says using the language of a physicist (actually quoting Kozyrev), "time is not a *scalar*" (by which the physicist means that time is not simply a *hypothetical* point, line or vector as we conceived of in that hypothetical geometry class I referred to a moment ago). Instead of a hypothetical *scalar*, time has substance and properties. Time *actively contributes* to what makes matter and energy what they are. Tesla's own words spell this out quite distinctly:

> During the succeeding two years (1893 and 1894) of intense concentration I was fortunate enough to make two far reaching discoveries. *The first was a dynamic theory of gravity*, which I have worked out in all details and hope to give to the world very soon. It explains the cause of this force and the motions of heavenly bodies under its influence so satisfactorily that it will put an end to the idle speculation and false conceptions, as that of curved space...
>
> Only the existence of a field of force can account for the motions of the bodies as observed and its assumption dispense with space curvature. All literature on this subject is futile and destined to oblivion. *So are all attempts to explain the working of the universe without recognizing the*

---

[20] Wilcock, David *The Source Field Investigations: The Hidden Science and Lost Civilizations behind the 2012 Prophecies*, New York: Dutton, 2011, p. 226; quoting Robert Youngson, Scientific Blunders: A Brief history of how wrong scientists can sometimes be. London: Constable & Robinson Publishing, 1998.

[21] Quoting from Albert Einstein, see http:///www.mathem.publro/proc/bsgp-10/0KOSTRO.PDF.

> *existence of the ether and the indispensable function it plays in the phenomena.*
>
> *...I consider myself the original discoverer of this truth, which can be expressed by the statement:* There is no energy in matter other than that received from the environment.
>
> *...It applies rigorously to molecules and atoms as well as to the largest heavenly bodies, and to all matter in the universe in any phase of its existence from its very first formation to its ultimate disintegration.* [22] [Emphasis in citation]

Such a notion regarding the essence of time is not only revolutionary; it begins to break down the wall between *nature* as we've known it, and *supernature* as we've felt obliged to characterize it – especially in theological discussions – to explain phenomena *beyond the observable world*.

In fact, several scientists suggest that time is what influences our universe to contain *spirals* – from the double-helix of DNA to the immense winding arms of galaxies. Why? Because time has a shape and it confers (imparts) that shape to matter if only to a slight extent. Time is like a whirlpool (*vortex* being the more proper physics term). Time can be *contorted* (energetically twisted out of shape). The issue of "where we are" and "when we are there" can only be explained by what is called *torsion physics*. We will discuss the concept of torsion more in a moment.

Furthermore, there is much else to say about how time is *structured*. The shaping of time is in fact compared by Farrell to a *crystalline latticework*. Just in case the reader isn't familiar with the structure of crystals, I will put a different analogy to work.

The universe at its most basic, most fundamental level consists of an invisible skeleton-like structure to which *everything* is "attached". Think of scaffolding filling an immense cathedral under renovation. Consider the scaffolding closely connected at its joints with pipes of very tiny diameter and length such that the scaffolding is able to fill

---

[22] Quoted from Farrell, Joseph P.'s book, *The Brotherhood of the Bell,* Kempton, Illinois, 2006, pp. 376-377; citing Tesla's words from O'Neill, John J., *Prodigal Genesis: The Life of Nikola Tesla* (Las Vegas: Brotherhood of Life, 1994), p. 250, emphasis added by Farrell.

every nook and cranny of this edifice. Additionally, this delicate scaffolding is so densely placed about… you can't see through it.

Having digested these notions, now conceive of the scaffolding as present everywhere throughout the cathedral as described so far, but add *invisibility*. It's there, you know it is, but you still can't see it. Finally, think of the cathedral as the universe and the scaffolding as the *aether* we call time which completely fills the Cosmos. Like this scaffolding in the cathedral, time is spread throughout the Cosmos with an invisible latticework to which all other particles attach.[23] This conception is actually quite fundamental to *what time is and does*. Granted, there is more to time than this underlying structure to reality – it has more properties, indeed it is "active" and not "passive" in affecting matter, i.e., anything that is must be attached to it. But first and foremost, *time* provides the universe's discrete skeleton.[24]

For thirty-three years Kozyrev experimented to test and prove this theory. These experiments, like the aether he sought to depict, were contentious. We read:

> What made them controversial was that Kozyrev viewed the spiraling patterns of nature and of life itself as a manifestation of time, and that as a consequence of this view, time itself was not a dimensionless "coordinate point" or a "scalar" as scientists and mathematicians would call it, but that time was itself a kind of physical force, and a very subtle one at that.[25]

One of the most provocative aspects of Kozyrev's theories, how-ever, is another stunning hypothesis that explains the behavior of fusion reactions: Farrell tells us that hydrogen bombs, like stars in the universe, generate far more energy that they should according to mathematical calculations. Therefore, additional energy must be included from some unexpected source.

---

[23] And I use the wording "other particles" intentionally because **time itself may be another particle**—an infinitesimally small—particle.

[24] Technically, what we find created is itself also a composition of time. Time is the raw material of the matter and energy that dwell within it.

[25] Ibid., p. 153.

Kozyrev's 1947 doctoral thesis made the audacious claim that the explanation for why stars and hydrogen bombs generate so much extra energy goes beyond the mere process of fusion – these massive reactions *are leeching (sucking) energy from time itself!* "In other words, Kozyrev had concluded that the geometry of local celestial space is a determinant in the energy output of fusion reactions, and that the latter, depending upon that geometry, will 'gate' now more, now less, energy into the reaction itself as a function of that geometry."[26]

Foreshadowing (actually falling on the heels of) the assertions of alchemists, the amount of energy depends upon the precise *status of time at that moment.* That is, time can fluctuate in its properties. Our common personal experience of feeling time moving faster or slower could be because it actually varies! *We could say, time doesn't convey the same timing all the time.* The quotation cited earlier referencing Einstein infers the same thing: The aether varies "from point to point."

The word used frequently by physicists like Farrell in depicting the physical nature of this model, so much so that it becomes a defining label for the model is, *Torsion*[27] *theory.* Farrell explains:

> Torsion is a dynamic and changing phenomenon, and not merely a static field, for if the basic idea of torsion is that it is related to a rotating system, [and] then it will be apparent that the universe is composed of rotating systems within other rotating systems, producing a continuously changing system with a changing flow of time. And if in addition those spinning systems are in turn emitting energy, such as a star pouring out its electromagnetic radiation from the fusion reactions within it, then that dynamic changes yet again, and with the constantly changing "spiraling and pleated" field of time, time itself takes on dynamic properties.
>
> Such complex, interlocked systems of rotation may be thought of as "knots" of space-time that are so intensely concentrated that they form the objects observed in the physical universe. As such, all systems are in fact "space-time machines," and since they "contain" space-time they

---

[26] Ibid., p. 155.

[27] Torsion is the distortion caused by applying torque in opposite directions to each end of an object—it is mechanical stress placed on an object by twisting it. A can of soda 'twisted back and forth' is an analogy that Farrell supplies to picture the nature of this process.

are not ultimately "constrained" by it, but rather, interact constantly, and in some cases, instantaneously, with it.

And this means that physics had to modify its mathematical modeling of time and space significantly.[28]

Consequently, with such a radically different approach to explaining the natural laws of the universe, "Kozyrev announced a wholesale assault on two of the foundations of modern physics and some of its hidden, and very counterintuitive, assumptions: Relativity and Quantum Mechanics."[29] That is to say, while Kozyrev offered alternative explanations for certain types of phenomena, he actually reinstated the common sense view that (1) time flows in one direction because it is in "time's nature" to do so[30], and (2) science can measure a particle's speed and location simultaneously without tampering with the results (we don't *force* any one of those properties to be contradicted just because we take a peek at it). Since more is involved in any such system which is being observed or measured—the *uncertainty principle* is not sacrosanct. In other words, what Heisenberg supposed, that scientific measurements are being made by scientists in closed systems (no other factors are involved to influence the measurements) turns out not to be a proper scientific way to think of these systems—any and all such systems being measured are *open systems* with multiple influences, not the least of which is *time itself*, since time is *not* a constant.

Further explanations for why this is so, I will leave to the curious reader who wishes to explore Farrell's explanations of Kozyrev's theory in more detail.[31] Consequently, "is it about time again for the aether?" Or should I just ask more to the point, "is time this aether?" In asking this question, a challenge surfaces because such language

---

[28] Ibid., p. 158.

[29] Ibid., p. 160.

[30] Wilcock challenges this, asserting that time flows into any particular point from all directions, swirling into an atom, a pyramid, or a galaxy.

[31] See Farrell's detail explanation on pages 161-165 of *The Philosopher's Stone*.

arouses contempt from the scientific community. Allow a Nobel Laureate to explain the situation I surface through his analysis of why we shouldn't use the scandalous term *aether*:

> It is ironic that Einstein's most creative work, the general theory of relativity, should boil down to conceptualizing space as a medium when his original premise [later, as we pointed out, he changed his mind] was that no such medium existed...
>
> The word "ether'" has extremely negative connotations in theoretical physics because of its past association with opposition to relativity. This is unfortunate because, stripped of these connotations; it rather nicely captures the way most physicists actually think about the vacuum. [*Vacuum* is the name most scientists would prefer in referring to this medium]. Relativity actually says nothing about the existence or nonexistence of matter pervading the universe, only that any such matter must have relativistic symmetry.
>
> It turns out that such matter exists. About the time relativity was becoming accepted, studies of radioactivity began showing that the empty vacuum of space had *spectroscopic structure similar to that of ordinary quantum solids and fluids* ["it" could be detected and measured – "it" existed]. Subsequent studies with large particle accelerators have now led us to understand that space is more like a piece of window glass than ideal Newtonian emptiness. It is filled with "stuff" that is normally transparent but can be made visible by hitting it sufficiently hard to knock out a part. The modern concept of the vacuum of space, confirmed every day by experiment, is a relativistic ether. But we do not call it this *because it is taboo* ["to do so" ... emphasis and comments added].

Why so taboo? Because once the aether is allowed into the discussion with the characteristics we've described here, we can't help but draw the conclusion the universe is that "fantastic reality" Pauwels and Bergier proposed. *Doors swing open to spiritually-oriented phenomenon where they were formerly shut tight.* No longer does nature rule out the possibility that fantastic events can and do happen. The nature of such events, such as the leeching of energy from time in fusion reactions, (or the transmutation of mercury into gold as we are about to discuss) is no longer impossible for a scientist to conceive. For those willing to examine the facts, it becomes a matter of documented history with top-shelf science to back it up.

Farrell provides a number of intriguing examples of how understanding exotic particles and a different explanation of space-time dimensions radically changes the science of physics and how the world works. His primary case study has to do with one David Hudson, an Arizona farmer, who in the 1980s discovered that alchemy is a fact and not merely a figment of an over-active occult imagination. For those who are aware of the unusual properties of the Ark of the Covenant, the discussion takes on special significance.

DAVID HUDSON'S GOLD STRIKE

Hudson spent several years and a small fortune, ultimately being rewarded with the discovery that his land was perhaps the richest in the world containing gold, platinum, iridium, and other precious metals—all *hiding* in plain sight—since it wasn't existing in its "natural" molecular manner. Through alchemical processes, initially administered quite by accident, Hudson struck it rich and in the process taught the world about *monatomic particles*, *super deformity*, *high spin rates*; and what certain chemicals and specific chemical processes—administered with extreme care—*beget a synthetically derived gold* (all loosely resembling the ancient quest of the alchemist!) Hudson didn't synthesize just any old gold, mind you—but precious metals of all sorts with exotic properties to boot. His story is nothing less than, shall we say, mesmerizing.

"You have," [said Hudson's metallurgist to him] "four to six ounces per ton of palladium, twelve to fourteen ounces per ton of platinum, a hundred fifty ounces per ton of osmium, two hundred fifty ounces per ton of ruthenium, six hundred ounces per ton of iridium..."[32] In other words, Hudson hadn't just struck gold, he had the richest set of precious metal deposits anywhere. It was lucky for him he had enough working capital to finance the investigation necessary to "prove his claim". "Working on this problem from 1983 until 1989, Hudson had employed 'one PhD chemist, three master chemists, (and) two technicians,' all working fulltime." Consequently, Hudson's quest to solve

---

[32] Ibid., p. 95.

various "strange" properties of the minerals on his land wound up making him the richest alchemist in history—with the possible exception of the Egyptian Pharaohs who may have understood alchemical properties and manufactured gold as many have speculated.[33]

Hudson's materials were capable of changing properties and becoming one of several different precious metals based upon exposures to intense heat for precise periods of time. One experience Farrell relates regarding the transmuted metal: by striking it with a hammer, the material emitted gamma rays (the blows triggered brilliantly bright mini-explosions). This property of the material was confirmed by experiments going on simultaneously at General Electric (GE). "To his surprise, Hudson had learned that the GE engineers had experienced the same unusual 'explosions' as well"[34] when they were working on devising new ways to create fuel cells with rhodium. "When our material was sent to them, [the GE engineers], the rhodium, as received, was analyzed to not have any rhodium in it. Yet when they mounted it on carbon in their fuel cell technology and ran the fuel cell for several weeks, it worked and it did what only rhodium would do…"[35]

But the most astounding property of the materials Hudson and his technicians were working with was yet to be explained: the weight increases and decreases of the material, *as if mass was being lost into thin air!* In a manner of speaking, this was exactly what they discovered. Taking the material through various heating and cooling states, the weight changed radically. This is not scientifically explainable—at least not according

---

[33] According to the episode, "Aliens and Temples of Gold", airing of August 18, 2011, from the *Ancient Aliens* program on the *History Channel,* the Kings surrounding Egypt asked the Pharaoh to share some of his gold inasmuch as it was "as common as the dust in Egypt." Several authors interviewed speculated that the Pyramids themselves may have been involved in generating gold from ancient technologies pre-dating the flood of Noah. Another fascinating anecdote: according to the broadcast, at the time of Sir Isaac Newton, his explorations into alchemy and the possibility of manufacturing gold from cheaper metals led the English Crown to *outlaw the practice of alchemy.* The royals didn't want the price of gold to fall if it turned out that gold could be manufactured from mercury or lead! Prices would have eroded if gold began to flood the market.

[34] Ibid., p. 95.

[35] Ibid., p. 95. When measured afterward, the rhodium was present.

to the rules of the "old physics". We read in Farrell's story quoting Hudson directly:

> We heated the material at one point two degrees per minute and cooled it at two degrees per minute. What we found is when you oxidize the material it weights 102%, when you hydro-reduce it, it weights 103%. So far so good. No problem. But when it [the material] turns snow white it only weights 56% of the beginning weight. Now that's impossible.
>
> If you put that on a silica test boat and you weighed it, it weights 56%. If you heat it to the point that it fuses into the glass, it turns black and all the weight returns. So the material hadn't volatized away. It was still there; it just couldn't be weighed any more. That's when everybody said this just isn't right; it can't be.
>
> Do you know that when we heated it and cooled it and heated it and cooled it and heated it and cooled it under helium or argon that when we cooled it, *it would weight [sic] three to four hundred percent of its beginning weight, and when we heated it, it would actually weigh less than nothing[?] If it wasn't in the pan, the pan would weigh more than the pan weighs when this stuff is in it.*[36] [Emphasis in original]

Just like in alchemy, as transmutations occur color changes are evident. What is most intriguing is that these color changes are consistent with the writings of alchemists from the time of the Renaissance. In other words, their experience as documented wasn't religious fiction or wishful thinking brought on by their pecuniary enthusiasm. Hudson's team was seeing atomic transmutations of elements causing elemental alterations to occur.

As a quick aside for readers sensitive to Judeo-Christian issues, given the properties of precious metals when treated with certain chemicals and massive heat, several mysteries of the Bible may find explanation. We know that Moses was an "Egyptian prince taught in all the ways of the Egyptians." (Acts 7:22)[37] Could it be that the Ark of the Covenant was influenced by alchemical knowledge learned by Moses in the court of Pharoah? Is this why the Ark was supposedly far lighter in weight

---

[36] Ibid., pp. 96-97.
[37] "And Moses was learned in all the wisdom of the Egyptians, and was mighty in words and in deeds." Acts 7:22.

than it should have been, having been constructed with gold veneers and a lid of solid gold? Could the exterior gold or possible internal contents containing this special gold—alchemically altered into what Hudson called "monatomic elements"—have been the explanation for why the Ark could apparently levitate?

The notion of how the strange dimensionality of this new reality works was published in March, 1989 in *Physical Review* by one of America's most highly regarded physicists: Dr. Hal Puthoff (who is associated with clairvoyance – aka *remote viewing* – and his work with America's so-called psychic spies at the Stanford Research Institute back in the 1970s). Hudson came across Puthoff's article which discussed a "strange white powder" that serves as a perfect superconductor. Puthoff explained that when it exists in this form, the material loses precisely four-ninths (4/9) of its weight, leaving five-ninths of its mass behind (5/9). It just so happens this fractional anomaly, once you do the simple math, equals 56% (55.5555…)[38] that Hudson and his team also observed.

> During their discussions, Puthoff told Hudson that "when this material only weights [sic] 56% of its true mass, you do realize that this material is actually *bending* space-time [?]" Such a material, Hudson noted, was what Puthoff "called exotic matter in his papers." Hudson had, in other words, literally stumbled across some of the exotic matter that forms so much of the quest of modern theoretical physics—not to mention mediaeval alchemy—and it was there right beneath his feet in the soil of his farm, and it was not really all that exotic at all. It was ordinary chemical elements, but in some sort of state not hitherto known. [Emphasis added]

But that wasn't all that Puthoff told him…

> If the mysterious white powder was indeed losing 44% of its mass, then said, Puthoff, "theoretically it should be withdrawing from these three dimensions… it should not even be in these three dimensions."[39]

---

[38] The repeating numeral "5" is certainly an occult number (think pentagram).

[39] Ibid., pp. 99-100

And with that mind-bender, we are ready for the next part of the story: uncovering how all of these exotic notions of the new reality were in fact discovered much earlier than Hudson or even Kozyrev; they were working hypotheses of the notorious *Third Reich,* in its vain attempt to develop a weapon of war that would make Germany victorious.

To that discussion we turn next.

THE THIRD REICH AND ITS TAKE ON THE NEW PHYSICS

A book was published before World War II (1933 to be precise) by Baron Rudolf von Sebottendorf, *Before Hitler Came*, which painted the occult picture of the Third Reich and so infuriated the Nazi's (perhaps for letting the cat out of the bag) that it was put on their index of prohibited publications. "Part of the reason may lie in the fact that, according to von Sebottendorf, the influence of the *Thulegesellschaft* [Thule Society] on the Nazi Party's formation and ritual was pervasive... [but] a more important reason for the ban on the work, however, must surely lie in the lists of its members that it contains, not only of prominent figures in the future Nazi State..." but identifying the linkage to a former inmate in an insane asylum who went on to become [Heinrich] Himmler's personal consultant on all things occult, *SS Brigadier General Karl Maria Wiligut*. "With recent scholarly publications, we are now also in a position to see the possible connection of alchemy to the SS' esoteric culture and its advanced physics projects via the various conceptions entertained by Wiligut, and passed on to Himmler."[40] Farrell even refers to Wiligut as equivalent to The Czar's "Rasputin". It is Wiligut's deep occult commitments and his teachings regarding alchemy that leads Farrell to agree that the relationship between secret societies in Germany did indeed run "directly to Hitler, and on that basis, such researchers [he mentions] often speculate that he [Hitler] was actually an initiate into one or more of these... esoteric societies and influences"[41] (e.g., *The Thule Society* and the *Order of the New Templars*).

---

[40] Ibid., pp. 248-249.
[41] Ibid., p. 250.

However, Farrell states it was *Wiligut's connection to Himmler of which we can be absolutely certain.*

> As such, it is less accurate to speak of an occult influence on the entire Nazi State, as it is to speak of an esoteric influence at the uppermost levels of the command structure of the SS. One is, so to speak, dealing with a Black Reich within the Reich, and at the uppermost reaches of the SS, with a very secret esoteric, and specifically *alchemical*, belief system.[42] [Emphasis added]

Farrell continues,

> While Wiligut's work to some extent paralleled that of the notorious SS *Ahnenerbedienst*, his work was essentially separate from that office. Wiligut worked *for Himmler personally*, whereas the *Ahnenerbe* was part of a much larger structure subject to more objective academic standards."

Wiligut was "instrumental in the selection and design of Himmler's infamous SS "Order Castle" at Wewelsburg, in the actual design of the SS ring, and the creation of SS rituals... he issued a 'steady stream of reports on esoteric matters of theology, history, and cosmology... for the most part *directly to Himmler.*[43] [Emphasis in original]

This connection and inspiration is important because it is Wiligut's alchemical influence on Himmler that motivated weapons research built upon a conception very similar to the "new reality" we've outlined in "alternative physics". Sounding almost like Pauwels and Bergier, Farrell states,

> In a certain sense, Himmler had willed into existence an entire government bureaucracy [*Ahnenerbe*] to do nothing but military studies of the esoteric, all under his personal control. This created an unusual if not unique first in modern history because for the first time in modern history a technologically and scientifically sophisticated great power was acknowledging, even if covertly, the existence of a very ancient Very High Civilization [a sophisticated pre-history civilization whose records are lost to us today] whose science it was intent upon recovering. Himmler had decreed, in effect, that the Third Reich was not only going

---

[42] Ibid., p. 250.
[43] Ibid., pp. 252-253.

to look for a "paleoancient Very High Civilization," for "Atlantis," but more importantly, for its science. [Emphasis in original][44]

One of the most striking comments we have to document this fact is from Wiligut to Himmler (June 17, 1936) which lands him right in the midst of a controversial discussion today amongst those who speculate that the *history of the solar system* is far more mysterious than we realize. It involves "worlds colliding" and peoples from another world tampering with the genetics of the human race for their own purposes. Wiligut asserts:

> Each of these evolutionary epochs which have occurred up to now were, according to the oral secret doctrine [an indirect reference to Madame Helena Petrovna Blavatsky's *Secret Doctrine*[45]], brought about by an enormous world-wide catastrophe culminated by unifications of our earth with one of the heavenly bodies attracted into its orbits.
>
> ...In the process, the remnants of humanity which remained on the earth assimilated with those who came "from heaven" (stars) to the "earth." This assimilation brought about similar intelligences and thus established a new humanity which instituted new racial types.[46]

Farrell goes on to explain that while aspects of these very ancient tales (today popularized by the works of the late Zechariah Sitchin derived from his translations of the Sumerian cuneiform tablets), reveals not just a cosmic war, but a war which used weapons "on a cosmic scale" involving rotating physics, i.e., the *vortical* or torsion nature of time described earlier. Farrell concludes this portion of his story with these words, "In short, of all the esotericists and occultists within the milieu of Himmler's SS, it is Wiligut himself who represents the best possible esoteric influence and basis for some of the SS' subsequent projects to reconstruct the technology of that physics."[47] This cosmic war requires a full discussion that Farrell pursues in several of his other books

---

[44] Ibid., p. 254.

[45] Blavatsky's Secret Doctrine was considered a fantasy to H.P. Lovecraft, but a fascinating one that served as a principal catalyst for his odd science fiction.

[46] Ibid., p. 257.

[47] Ibid., p. 258.

which falls outside this study's scope.[48] While it cannot be developed in all its depth, it will be touched on further in the next chapter.

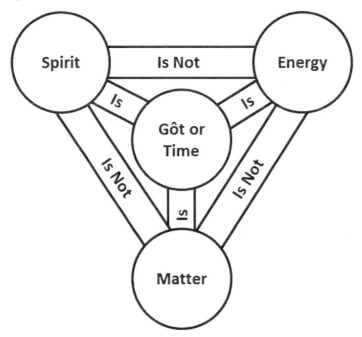

Figure 42 – WILIGUT'S VERSION OF THE TRUINE STONE

While we don't delve into that subject further here, Farrell is referring to the *Nazi Bell*, a famed *wonder weapon* based upon its ability to generate a torsion field that could leech massive energy from "the medium", i.e., time. In effect, the Nazi Bell would surpass the power and *practical* use (if I may be allowed to call it that) of the atomic bomb by turning time itself into a radiation-free hyper-explosive. (The Bell could make the manipulation of *time* a weapon of war by incorporating an *alchemical* process by opposing hyper-rotating cylinders to steal energy from time and transform it into a mega-bomb.)

Farrell provides a number of direct quotes from Wiligut that speak of "the primal twist", a "Rotating Eye" (*Drehauge*)[49] when speaking of the flow of Matter-Energy-Spirit. "Gôt is eternal—as Time, Space, Energy

---

[48] See Farrell's catalogue at en.wikipedia.org/wiki/Joseph_P._Farrell.
[49] Cited by Farrell, ibid. p. 258, from *Flowers and Moynihan, The Secret King*, p. 70

and Matter in his circulating current."[50] Therefore, Farrell concludes a fascinating axiom to a revised Cosmos we find appealing albeit partial:

> The physics jargon employed – "circulating current"—suggests that he has in mind precisely a physics metaphor or interpretation for the whole scheme, for a circulating current suggests precisely time as the fundamental component of his "primal twist" ... By invoking a physics terminology with his reference to "circulating current," Wiligut is implying that this whole alchemical-theological theme is at root based in physics and not metaphysics[51]. Thus, as he avers, this "Gôt" is "beyond the concepts of good and evil." In a Europe already concussed with loss of faith and a confidence in Christian institutions and morality, such a statement could only have disastrous moral consequences, for the whole mistaken thematic identity of this physical process with the Christian Trinity and therefore with the whole edifice of Christian morality inevitably led to a similar mistake in ethics: if they were based only in a physical process neither good nor evil, then there could be no ultimately good or evil action, only actions now more, now less, in harmony with the process itself. With Wiligut's "triune Stone" the way was open for the genocidal alchemical transmutation of man himself.[52]

In other words, by associating God with *time* (the "new reality"), good and evil as existing in truth in the Cosmos becomes discredited because there no personal God exists to make them objectively real and not mere subjective perspectives convenient to finite creatures or their societies. To Wiligut, God is impersonal. God doesn't transcend the natural universe, *albeit a mysterious nature.*

Consequently, the Nazi aspirations to transform man into the *Übermensch* (or annihilate the Jews for failing to measure up to the capability of "the new man" because of their inferior "blood") are perfectly consistent with the reality of nature. According to Wiligut, *Gôt* resides in

---

[50] Cited by Farrell, ibid. p. 258, from *Flowers and Moynihan, The Secret King*, p. 79, from Karl Maria Wiligut, "The Nine Commandments of Gôt," signed manuscript.

[51] *Metaphysics* is defined by the *Encarta Dictionary* as "the branch of philosophy concerned with the study of the nature of being and beings, existence, time and space, and causality." However, in this context Farrell is comparing *physics to nature* and *metaphysics to supernature*, inasmuch as *Meta* means "beyond" or "surpassing". Adopting this distinction (incorporating the same reference to metaphysics Farrell asserts), will be supportive to our discussion as well.

[52] Ibid., pp. 262-263.

the natural or physical world—not the metaphysical (or supernatural) realm.

Note: this alchemy—transmuting *a personal God into an impersonal force*—eradicates any element of morality from the Cosmos. Morality is arbitrary and not absolute. It is subject to variation based upon what the majority or the dictator declares to be moral.

The *triune Stone* mentioned comprises a pictogram Farrell presents not just for Wiligut (but for St. Augustine and others) to portray how various true or would-be scholars have understood the fundamental nature of God and His analogue in the world of matter, energy, space, and time. Wiligut's version was originally drawn by Farrell (and reproduced by me) in the figure earlier.

Farrell indicates that there are other statements Wiligut makes that provide the ideological connection between his alchemical physics and the "highly classified secret weapons research project, *The Bell*; statements that indicate that, in part, these esoteric views might have formed part of the rationalization for the project, at least, as far as the unscientific leadership of the SS was concerned."[53] Wiligut discussed research that was completed prior to 1933 involving the separation of two rotating fields—going in opposite directions—which was the basis for *The Bell* project. "Wiligut is implying by means of his reference to the swastika and the torsion-based physics it symbolized, that the fabric of space-time has a spin orientation."[54]

Figure 43 – THE INSIGNIA OF THE THULE SOCIETY

---

[53] Ibid., p. 263.

[54] Ibid., p. 264. The article by Wiligut referenced is entitled "Ancient Family Crest of the House of Wiligut" published in the magazine *Hagal* in 1933.

Indeed, the contention is that *the symbol of the swastika itself reinforces this notion*! When we evaluate the swastika as illustrated in the 1919 insignia of the *Thule Society* (see figure above), the 'spin' nature of the emblem is strongly suggested.

Farrell provides compelling additional evidence for the alchemical connections of Wiligut, Himmler, and secret weapon research of the Nazis based upon the *torsion* conception of time. Interested readers are encouraged to investigate his evidence further.

NORMAL VS. PARANORMAL: WHY THE DISTINCTION?

The point of this lesson in "revised" physics and how it relates to my thesis is simply this: what many formerly considered an example of magic—notably *identifying the philosopher's stone* (as we've learned from Joseph Farrell's exploration into comparing the ancient art of alchemy with the new insights of exotic particles), is now radiating (pun intended) astounding new insights into *the discoveries of revised atomic theory*. Alchemy, the ancient art of turning lead, mercury, or other non-precious metals into gold, has been proven to be much more than a magic trick of a showman or a genuine act of wizardry. It comprises an authentic science if conducted in light of what we have discussed here.

It turns out to be the case that *the alchemist is a chemist after all*—we can now see the alchemist simply practices a very advanced and seldom understood form of this science. All talk of hocus pocus and personal transformation aside (core aspects of the alchemical folklore), along with understanding the nature of exotic particles in the genuinely 'natural' universe, underscores how postulating supernatural causes and effects for many phenomenon is actually just another way of saying *"we now seek to understand the science of the unusual"* (but not the impossible). The universe is very different than what we once thought. *Its behavior can be so strange it almost seems supernatural.* But upon closer inspection and a better conceptualization, this aspect of 'supernatural' is still natural—it's just our empirical understanding has progressed, taking into account natural laws we didn't comprehend before.

Consequently, this enlightened awareness doesn't disparage science in the least; and furthermore, it certainly doesn't dismiss science. If anything, it indicates that science stands as a valid and necessary human activity. The point to be made is this: science *must maintain humility*. As smart as we think we are, the final word hasn't been spoken. We still have much to learn. Speaking somewhat metaphorically here, the instruments we use through our scientific inquiries—and even more importantly, how we interpret their results—need to be considered cautiously and our axioms considered *provisional*; that is, they must always remain open for the possibility of further mid-course correction. The bottom line: we will continue traveling down the scientific road of discovery until time itself (for those adopting a Judeo-Christian perspective at least) arrives at its conclusion. Many phenomena that might have caused us to think we were dealing with the supernatural—having no measurable traces of physical, chemical, or electromagnetic forces—may still be explained by an enhanced understanding of the universe and its *natural* laws. Just because we can't fully articulate what those laws are today doesn't mean they do not exist.

However, here we must assert a strongly worded BUT. This improved understanding of the organic, albeit misunderstood forces of nature, doesn't imply that all that is *is nature*. In other words, *a distinct supernatural or paranormal may still exist*. For those of faith, *it does still exist*.

Certain phenomena for people of faith are still yet to be explained. An "expanded nature" must be supplemented by powers, notably personal in character, that can't be reduced to nothing more than impersonal "forces" or "fields". They can be persons, even if they are superhuman, with capacities that far exceed our own; entities occupying dimensions superseding our everyday experience.

CONCLUSION: POWERS BEYOND THE SUPERNATURAL

The Nazis delved into both sides of this new reality, or more precisely, an uncommon *understanding* of reality. The first portion of the new reality was nature transformed by a distinct understanding of space-time and how to manipulate the medium (the aether) to create true and deadly weapons of mass destruction that likely exists and remain "under wraps". But the portion of reality, what Christians and Jews distinguish to be eternity, still remains distinct. Spatial language can throw us off. This reality may "underlie" the reality perceptible to our senses. This reality may "transcend" our senses. And this reality may not be separate from but rather inclusive of that which we can touch, taste, feel, hear, and see.

In this other aspect of the "new reality" (new age author David Wilcock labels it time-space in distinction to space-time), we encounter and engage *personal* forces. Our encounters of these forces could be, to use the words of spiritual warfare,[55] "bound or loosed" at the will of the magician (or more aptly, the spiritual warrior). To be clear on this point, these forces, be they *demonic* (or to use Jung's terms, *archetypal)*, still exist in traditional Judeo-Christian cosmology *despite the new discoveries or ways of thinking about nature we've just discussed.*

As author Trevor Ravenscroft (author of *The Spear of Destiny*) and author Joseph P. Farrell to some extent document, Hitler became a ritual magician through the tutoring of one Dietrich Eckart. The occult mediums of the Third Reich took actions based upon the powers that magicians purportedly possessed. They may have connected these forces with the ethereal *Vril* of Sir Edward Bulwer-Lytton and The Vril Society of Maria Orsic. But there remains little doubt they also believed and interacted with *personal* forces of evil. These forces transcend the "fantastic reality" of Pauwels and Bergier. *They substantiate the cosmology of the occult and the Bible (to the extent we give them credence).* In this sense, **the Cosmos constitutes a personal reality inhabited by spiritual beings:** demons, angels, and perhaps those that have passed on from this life into the next. But most importantly, the principal deity, the Elohim of Elohim, remains *Jehovah God*. Ironically, the Nazis concluded that this God was an *impersonal*

---

[55] Spiritual warfare is a phrase that Christians use to reflect the process of challenging spiritual forces, which may be visible or audible during the confrontation. Exorcism is the most radical and notorious type of 'warfare'—but prayer, while less audacious, may nevertheless comprise a primary weapon of choice.

force—despite the fact they still knowingly engaged with *personal demons*! However illogical this may seem, the facts state this plainly.

While pop Spiritism often makes use of the "new reality" *of the first type* (the *fantastic reality* of Pauwels and Bergier) to substantiate its occult beliefs and invoke ritual magic, it doesn't logically follow that the Spiritism to which they pay heed *comprises an impersonal force,* despite the fact it is how they choose to characterize *the medium.* Such a supposition amounts leads to the dead-end where the detour taken by these advocates of ancient wisdom and the commonly used label, the New Age, takes them.[56] As we will discuss, occult enthusiasts may depict the new reality *pantheistically*. Moreover, they often describe it as humankind's super-consciousness that transcends individual minds and resides in the supernatural world—but this concept of consciousness, *according to them*, belies the power unique to humanity which we can and should master. Jung's Collective Unconscious is the classic label affixed to this likely reality. In their estimation, by growing increasingly aware of this reality – no matter how ethereal or elusive – we become spiritually adept and advance beyond others.

However, in the final analysis, their instruction is quite deceptive. For while the new reality provides fantastic new explanations for why nature works the way it does, it offers no explanation for why there are genuine spiritual forces which still dwell beyond nature – no matter how fanciful – that can be influenced by the technical know-how of the magician (or for that matter. *the prayer of the godly*). For as many authors in this genre indicate (including Farrell when mentioning the infamous aliens who purportedly abduct humans), such extraterrestrial origin is suspicious for many scientific reasons; but mostly because it challenges the veracity of the assertion the "ETs" are otherworldly beneficent beings essentially demonstrated by the evil behavior they exhibit. Indeed, it is this factor (that these forces *can be and are evil*), which argues so convincingly that *the cosmology of the Bible is the most persuasive worldview offering explanation to why such evil exists*. If authentic good and evil inhabit the universe (in other

---

[56] If the book of Revelation be true, as Christians assert: a second death.

words, good and evil are real, they aren't just a sentimental verdict or metaphoric judgment), *impersonal forces* are no longer apt explanations for the essence of these powers. At its most fundamental level, going beyond neutrons, gravitons, and any other exotic particle, *reality is personal*. Saying it another way, **the Cosmos consists of entities that are moral agents chiefly because God is a moral agent and he holds all the other moral agents of the Cosmos accountable for what they do. Righteousness is real. So is evil. But it is real because both are incarnated in personages – be they heavenly or human.**

As we open up to this *other realm of the new reality* and the power quest of individuals who seek to master it – some with only beneficent motives and others maleficent motives, the search for understanding the Cosmos begs the question of meaning and the answer to that question begins with personhood: human beings and "gods". At that top of that spectrum of personalities is the God of Gods, the unique God in three persons, Yahweh. His existence and revelation to humankind declares that the metaphysical – the "really real" exists; it is thoroughly moral (good and evil exist not dualistically but truly), and personal beings inhabit it who embody this reality. If so, as said above, the most fundamental reality of all – God as made known to humanity through Jesus Christ – must be *personal and ultimately* the basis for all that is; what theologians have called "the ground of being". Speaking (as the philosophers say) *ontologically*, God is the foundation of all that is.

Consequently, at the risk of overstating it yet one more time, that which is *good and evil* truly exists in the Cosmos. Like alchemy was once believed to be, these attributes are much more than a figment of our spiritual imagination. Speaking literally and not metaphorically, *personhood* lies at the core of what exists. It comprises the very essence of the universe; it constitutes the warp and woof of life. Personal beings are the pinnacle of the Cosmos. Personality constitutes the most important aspect of the universe, for it reflects the very nature of God Himself.

MAN-MACHINE HYBRID: THE NETWORKED CHIMERA AWAKENS
Gonzo Shimura with Doug Woodward

"For the scientist who has lived by his faith in the power of reason, the story ends like a bad dream. He has scaled the mountain of ignorance; he is about to conquer the highest peak; as he pulls himself over the final rock, he is greeted by a band of theologians who have been sitting there for centuries."

Robert Jastrow
The Enchanted Loom: Mind in the Universe

INTRODUCTION

FOR CENTURIES, HUMANKIND AFFIRMED THAT THE UNIVERSE AND ITS SENTIENT CREATURES WERE DESIGNED BY A VASTLY MORE INTELLIGENT AND MORE DEEPLY SENTIENT CREATOR. BUT AS THE Enlightenment advanced in the old world as well as in the new, academic institutions began emphasizing human reason more than biblical revelation to explain the metaphysical, i.e., the foundations of ultimate reality. In short, intellectuals from many disciplines sought to explain the universe within the boundaries of natural law, without resort to a creator. Steadily, reference to the supernatural to express "what is" gave way exclusively to natural explanations to comprehend what exists and how it operates.

In the Western world, science became independent of metaphysics in guiding humanity's thinking. While a "deistic" God was allowed by intellectuals in most fields of study (the Founding Fathers in America saw God this way), the anthropomorphic aspects of the biblical God (most notably a God-man in the person of Jesus Christ) were rejected.

In our day, ironically, when trying to discern and manipulate natural law, metaphysics has revived. And now, energized with the arcane discoveries of today's science, the supernatural worldview of the Bible no longer seems so outlandish or passé.

For centuries, transportation and communication methodologies were limited to riding on the backs of animals, using our own two feet, or by boat if one desired to traverse beyond the great deep. Communication was largely done by word of mouth, or by written materials that needed transporting via these same lethargic means. But the twentieth century changed everything. Automobiles, trains, and planes made transportation faster and more efficient. And while communication by electronic means began in the nineteenth century, it wasn't until the twentieth century that the telephone, television, and radio become household norms. This dramatic phase of modernization in the twentieth century is considered by many to be the third phase of the Industrial Revolution.[1]

As we charge towards what is called the "Fourth Industrial Revolution," we are faced with even more radical changes to the infrastructure of civilization. Integration of synthetic biology, nanotechnology, quantum sciences, particle physics, deep space exploration, deep sea exploration, Artificial Intelligence, clean energy, virtual/digital realities, advanced information technology, and many other subjects of inquiry have already begun transforming society. Through this revolution, we are being promised a new "Golden Age" with transportation going celestial and communication going quantum.[2] Indeed, natural law has reached the boundaries of empirical reality and now seeks to transcend them. At this very

---

[1] Historians suggest that the First Industrial Revolution occurred between the 1790s and the 1840s, where basic machines harnessed the power of wind, horse, man, and water in factory settings where textiles and fabrics began to be produced in faster, more efficient means. The Second Industrial Revolution occurred between 1870-1910. As a means to rebuild after the Civil War, large projects involving railroads and basic control of electricity were achieved. This is also the era that big bank moguls of the day (Rockefeller, Carnegie, Vanderbilt, J.P. Morgan and Edison) considered as "robber barons" started to take over social construction. The Third Industrial Revolution is thought to have begun in the 1970s where information technology started to rapidly increase and change the way the world operated. This era is sometimes referred to as the "Information age."

[2] Many occultist philosophies from the "Golden Dawn", Rosicrucians, Freemasons, Theosophists, Kabbalists and many others, believe that mankind is destined to reunite

moment, scientific inquiry knocks on the doors of invisible dimensions that theologians have consistently proclaimed for millennia. With advance of knowledge finally hitting the barriers imposed by natural law frustrating further growth (as understood by the "Enlightenment Mind"), many scientists look to mystical religion and occult mythos, searching for keys to another golden era of humanity. But do these occult visions and practices promise universal benefits for mankind? Or do these mystical tales yield secrets only to an initiated elite?

The authors of this book assert that what lies beyond spiritual doorways proves to be neither benign nor beneficial. Many passages from the Bible warn humankind to leave dimensional portals alone since powers lurk behind them that threaten our existence. Within these dimensions await deceptions of maleficent spiritual enemies to our souls.

> *For we do not wrestle against flesh and blood, but against the rulers, against the authorities, against the cosmic powers over this present darkness, against the spiritual forces of evil in the heavenly places.* (Ephesians 6:12)

The authors of this book feel duty-bound to describe spiritual realities which exist and conditions that surely produce nefarious outcomes. Our caution isn't meant to be judgmental to merely justify ourselves. The caution isn't directed at one particular person or group. We can lump all spiritual pursuits together and assert that any path which seeks spiritual reality without acknowledging the God of the Bible, Yahweh, and His Only Son, Jesus Christ, will ultimately deceive the explorer into adopting a synthetic (man-made) substitute. The true solution that brings resolution to all such matters of the spirit involves what Christians label "Christ's work on the cross".

We cannot discern if any given person's beliefs adopt His work for salvific affect. Only He knows who are His. This includes those whose pursuits from our vantage point border on occultism. Our calling? To point out the workings of the enemy through his deception foisted on

---

heaven and earth in a return to the pre-historic Golden Age that many believe hosted a civilization perhaps more advanced than is ours today.

humanity, while presuming God maintains complete control, and within Jesus Christ lies all truth.

THE NEW AGE OF THE INDUSTRIAL REVOLUTION

If history has any bearing on the deployment of these pseudo-spiritual discoveries (perhaps "practically paranormal" would be a better choice of words), then no doubt dark spiritual intelligences will conspire to control them. In the initial advent of the Industrial Revolution, the manufacturing of goods changed radically. This brought about new methods to game commercial systems and devise more arcane means to finance business and government. It was during this era, that the Rothschild family infamously conspired to take over the British government by manipulating its securities market.[3]

Central banking, how money was created, and fiat-based currencies dramatically evolved. And with control of the economy by "banksters" instead of State treasuries, the building of transportation and communication infrastructure relied upon debt issued to governments created out of thin air rather than money issued by the State backed by commodities like gold and silver. For centuries precious metals were the "backing" of paper money. For the United States, however (and *de facto* for the world due to American hegemony), this changed radically under President Richard Nixon. He took us off "the gold standard".

In 1971, oil ("black gold") became the preferred global commodity as a means to back the U.S. Dollar instead of gold or silver, and was shortly the world's reserve currency (displacing the British Pound; hence, why it's called "the Petro-Dollar"). It's a widely accepted bit of conspiratorial history that the oil industry has suppressed more efficient, cleaner, and readily available forms of energy for transportation

---

[3] The House of Rothschild used strategically placed couriers to learn the outcome of the Battle of Waterloo three days before anyone else in England. By spreading rumors that the British had lost, their stock markets braced for the loss. In the process, the Rothschild's bought all the bonds being rapidly sold off, and by the time the real news arrived, it was too late: The House of Rothschild effectively owned England. *Ferguson, Niall., The House of Rothschild: Volume 1: Money's Prophets: 1798-1848, https://goo.gl/CPxajJ*

technologies. In part this is to keep the price of oil high for the sake of making oil barons rich – but it also keeps the backing of currencies stable so that how much or how little inflation takes place, and remains under the control of the banksters.

We see a similar shift when one looks into how deconstructing railroads was accomplished in conjunction with the boom of the automobile industry.[4] The U.S. proclaims its adherence to "free market capitalism". Indeed, it is a proven way to build great wealth and maintain a middle class. It is the most effective way to generate the broadest amount of wealth. However, if the "system" is run by big corporations directing legislation (via lobbyists to achieve their ends), then free markets are no longer free. Instead, America's system becomes a red, white, and blue form of fascism.

But lately, the disparity between rich and poor has become a vast chasm. This leads us to ask, "Aren't markets just well-managed monopolies dominated by the wealthy?" Said beloved Christian author C.S. Lewis, "That is the key to history. Terrific energy is expended – civilizations are built up – excellent institutions devised; but each time something goes wrong. Some fatal flaw always brings the selfish and the cruel people to the top and it all slides back into misery and ruin."[5] As Lewis hints, civilizations have always been subject to violent flux as well as the schemes of the rich to get richer. But vast increases in levels of computational power amplify the next wave of "expended energy" in the fourth Industrial Revolution. These technological breakthroughs portend opportunities for even greater wealth and inequality between rich and poor.

To be sure, the foothills of this digital revolution have already been scaled. We needn't reconnoiter the many trials which ascend to the peaks before us, let alone trace the innumerable rabbit trails that would

---

[4] *https://www.corbettreport.com/episode-310-rise-of-the-oiligarchs/*
[5] C.S. Lewis, *Mere Christianity* (New York: Macmillan, 1952), pp 53-54.

distract us from our plotted course for this chapter. Still, it's worth mentioning a couple of examples just to make our message clear.

First, consider Tesla Motors. It is no mere coincidence that the modern progressive transportation company bears the name of a man who not only allegedly discovered ways to tap into so-called free energy over 100 years ago, but had run-ins with the banking elite. The conflict was easily explained: the banksters and corporate powers would have suffered untold financial loss as a result of deploying his discoveries. (Therefore, no mystery exists surrounding why Tesla's mountain of revolutionary technologies was confiscated upon his death).

The face and CEO of the modern corporate entity serving as his namesake, billionaire Elon Musk, now serves as an icon of innovation for an adoring public. For those who grasp the rise and fall of civilizations, Musk symbolizes goals and aspirations reflecting occultist sentiments longing for the next Golden Era. "We want to open up space for humanity, and in order to do that, space must be affordable… If humanity is to become multi-planetary, the fundamental breakthrough that needs to occur in rocketry is a rapidly and completely reusable rocket."[6]

The innovations Musk pushes for in the realm of rocketry directly connect to the occult. Jack Parsons, co-founder of the Jet Propulsion Laboratory (JPL), was one of America's major contributors to rocket development devising solid rocket fuel, a much safer means to propulsion than the liquid rocket fuel utilized before. Parsons was an enthusiastic follower of notorious English occultist Aleister Crowley, who referred to himself as "The Great Beast 666." Parsons became the head of the Crowley's occult religion employing "sexual magick" and dark perversions; specifically, the *Ordo Templi Orientis* (OTO) in Pasadena, California. Crowley apparently considered Parsons a nuisance, but the impact of Parsons on the world of science fiction, UFOs, and occult occurrences in America remains indisputable. His association with founder of Scientology, L. Ron Hubbard, makes for

---

[6] http://www.techinsider.io/elon-musk-future-quotes-2016-2

one of the most bizarre tales of military intrigue and marital infidelity ever told in the United States.[7] But we won't go there in this book.

So it is no surprise to learn that Elon Musk, and his private aerospace manufacturing and space transport service company, *Space-X*, have been testing a robotic space capsule named "Dragon." According to Musk, the name "Dragon" came from the 1963 song. "Puff the Magic Dragon" by folk singers *Peter, Paul and Mary*. Since this author doesn't believe in coincidences of such magnitude, and considering the spiritual nature of influence behind such prominent corporate decisions, naming the capsule "Dragon" might very well have resulted from whispers of the Fallen One. Therefore, when Space-X tweeted "Planning to send Dragon to Mars as soon as 2018. Red Dragons will inform overall Mars architecture…"[8] or when we read headlines that say "Dragon Returns to Earth…"[9] we can't help but be reminded of what we are told in the Scriptures:

> *And the great dragon was cast out (thrown down), that old serpent, called the Devil, and Satan, which deceiveth the whole world: he was cast out into the earth, and his angels were cast out with him.* (Revelation 12:9)

While the first example involved "going celestial" in humanity's quest to extend its domain to the Cosmos beyond our planet, the second example deals with "going quantum" – going the other direction – into the ultimately tiny. And yet it has equal weight in the exploration of the Cosmos and even greater meaning to the destiny of humanity. In essence, we have gone large and now we will go small. These two areas of inquiry, paradoxically, are actually folding back onto themselves.

---

[7] See Woodward's recounting of this material in his book, *Lying Wonders of the Red Planet (2012),* Oklahoma City: Faith Happens, chapter 3, "A Rocket Scientist and Occult Priest", pp. 41-51.

[8] https://twitter.com/spacex/status/725351354537906176

[9] https://blogs.nasa.gov/spacestation/2016/05/11/dragon-returns-to-earth-in-pacific-splashdown/

THE CHIMERA COLOSSUS COMES TO LIFE

> *"You talk as if a god had made the Machine," cried the other. "I believe that you pray to it when you are unhappy. Men made it, do not forget that. Great men, but men. The Machine is much, but not everything."*
>
> E.M. Forster, *The Machine Stops*, 1909

Over 100 years ago, author E.M. Forster penned a short story that could be deemed prophetic literature. "The Machine Stops" depicts a dystopian future where humanity is no longer able to live on the surface of the earth, but must relocate to individual cells underground where an omnipotent global machine provides all of humankind's basic physical and spiritual needs. What is most remarkable about Forster's depiction of this subterranean world? *Its communication technology.*

It eerily resembles "video chatting" via the Internet – much as we experience on Skype and FaceTime today. Furthermore, in Forster's world, those who do not accept the mechanized deity are considered unfit for society and ultimately banished from the underground civilization, expelled onto the uninhabitable surface of the earth. (In passing, we should mention that the discerning Forster touches on a number of other popular current-day "fringe" topics like underground bases, climate change, and unfamiliar means of air travel.) In the apocalyptic closing scenes of the short story, the god-like machine collapses and the characters realize that their connection to nature was

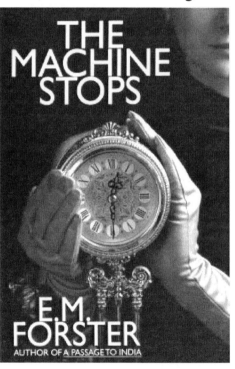

Figure 44 – THE HIGHLY REGARDED WORK BY E.M. FORSTER, THE MACHINE STOPS

the "purpose of life" all along. Foreshadowing the *Georgia Guidestones*,[10] they attempt to extend a warning to surface dwellers and future rebuilders of human civilizations not to repeat the mistakes made by the builders of the Machine.

While the inclinations of Forster are disturbingly portentous, there is an even older text I take to be of supernatural origin that supplies similar warning. It predicts an evil personage that:

> *Deceiveth them that dwell on the earth by the means of those miracles which he had power to do in the sight of the beast; saying to them that dwell on the earth, that they should make an image to the beast, which had the wound by a sword, and did live. And he had power to give life unto the image of the beast, that the image of the beast should both speak, and cause that as many as would not worship the image of the beast should be killed.* (Revelation 13:14-15)

The backstory to this passage: the arrival of the infamous first beast who emerges from the sea (a topic we will touch on later) that incurs a mortal head wound which is miraculously healed. This preternatural healing, which some commentators believe constitutes a counterfeit resurrection scenario of the Antichrist, is prelude to the beast that emerges from the earth. This second beast deceives mankind into constructing the image of the first beast, much like the people built the Machine in Forster's dystopic future. And as Forster's machine identified with attributes of God, the second beast gives "life unto the *image* of the (first) beast."

Figure 45 - THE COMPANY WHICH CREATED SKYNET IN THE *TERMINATOR* MOVIE SERIES.

---

[10] See https://en.wikipedia.org/wiki/Georgia_Guidestones for details.

As of this writing on the cusp of the Fourth Industrial Revolution, peering ahead at the massive social change which awaits us, it's inconceivable to overlook the relevance of this dreadful vision dominating the Book of Revelation. What Forster foresaw over 100 years ago, we witness now. And what the Scriptures prophesied thousands of years ago match the state of society today and technology's apparent trajectory which looms before us causing the observant and informed considerable trepidation. (Even astute movie goers recognize the portentous character of the Terminator movie series given that the Terminators' all machine network, *Skynet*, originally built by the fictional Cyberdyne Systems, sought to destroy humanity after it became "self-aware").

An image seemingly of an inanimate object, projects lifelike qualities. It takes little imagination to compare this image to the contemporary development of Artificial Intelligence (A.I.). Likewise, failure to worship the image, as with Forster's Machine, sentences any such rebel to death. Whether it does or doesn't achieve actual sentience, this image will possess God-like attributes (mimicking omnipotence, omnipresence, and the power to give and take life.) The questions to consider are not whether the machine displays genuine intelligence (nor whether it acquires a soul at some point); but rather, "What are the spiritual implications of its presence, the fact that it exists at all, and what it will demand from its subjects? How we might respond given we seem to be living the times that not only Forster saw, but also the future the Bible prophesied?" And paramount question, "What does this technological Frankenstein want with OUR souls?"

Today, we cannot imagine a world without computing machines. In fact, this very document would not have been created without modern computing. Yet, even in 1909, Forster's story warned us that humans would grow dangerously dependent on technology. His idea was not unprecedented. H.G. Wells, Forster's contemporary, had published *A Modern Utopia*, published four years earlier in 1905. Wells also foresaw a civilization reliant upon technology. Perhaps unpredictably, both Forster and Wells tapped into the frightening visions of the biblical *eschaton*, transforming the demonic quests of the two beasts into "cold, calculating" machines that would possess and then overwhelm their makers. They took opposing sides on the ethical (indeed spiritual) spectrum – one which embraced this coming "Technium" (see below) to its exalted conclusion, and one who rejected it outright fearing its consequences. In a world where machines far outstrip humanity, inevitably a new religion will be hatched like a serpent. Should the eschatology of the Bible prove true (and we assert it will), those who refuse to bow to the image and reject the *abomination*, the "pseudo-life" the all-powerful machine begets, will be excluded from society and soon thereafter martyred doubtless in ritual manner.

THE METAPHYSICS OF THE TECHNIUM

Clearly, Forster's outlook on the relationship between mankind and technology seems precognitive. However, he did not foresee the merging of *metaphysical understandings with the knowledgebase of natural law*. While unforeseen to Forster, our world has boldly redrawn the line (better yet, thinning the line almost to non-existence) distinguishing between religion from science, ignoring the rebuke of the Enlightenment in its outlook on the realm of the supernatural.

Commonly, denizens of the post-modern world declare a given person "spiritual" without reference to his or her being particularly "religious." "Being religious" now conveys constricted thinking and attenuated behavior, personal immaturity, reliance on failed institutions, and perhaps ironically, strongly inclined to moralizing which is itself deemed immoral. Nevertheless, by ascribing spirituality as an admirable attribute – meager as it may seem to devoted Christians – our culture admits humanity's quest for transcendent qualities in its existence. Life retains a certain sanctity for us all, even the most ardent atheist. Likewise, the nihilist must grudgingly admit that he or she must identify purpose too, however illusory, if our lives are to be satisfying. Therefore, so it is that human civilization will confront a profound paradox when we encounter "intelligence" in the *Technium*, the *Machine cast as the image of the beast of Revelation 13.*[11]

The more science advances the more it uncovers evidence for realities surpassing the purely empirical (that is, transcending the five senses).[12] Scientific perspectives are no longer divorced from magical thinking. As Arthur C. Clarke said, *"Any sufficiently advanced technology is indistinguishable from Magic."*

The word *Technium*, coined by co-creator of *Wired Magazine* Kevin Kelly, supposes *collective interrelationships* between all machines on the planet. According to Kelly, this network of interworking machines forms a life of sorts. An everyday example Kelly uses: a fountainpen. The pen itself is a technological tool in its own right, but to come into being, it required other technologies from many different disciplines. The independent development of ink, metal, plastic, and ball bearings all journeyed to an intersection where a fountain pen was formed. While the pen may be mightier than the sword, it does not possess self-awareness. Nevertheless, Kelly employs this analogue to illustrate the concept of his Technium.

Therefore, the Technium is the ultimate network of all mechanized technologies combined in a supportive fashion, supplementing one other, which equals more than the sum of its parts. The Technium even exhibits life-like behaviors in the same way that a single neuron in your brain doesn't "think", but together the *network of neurons* acts as a fused *collective* to form an idea.

Consequently, we see the network of all the world's technology – past as well as present – forming a virtual ecosystem which appears to possess its own urges and ambitions. Like any kind of a system, it has certain biases, and those biases derive from the system. It doesn't really matter whether particular humans are present to influence it. (It's okay for the

---

[11] I will capitalize Technium throughout conveying it possesses deity – not because it actually does, but because the cultural thinking about it assumes that it is a god.

[12] In philosophy, this view is known as logical positivism or simply positivism. It is more commonly referred to as "naturalism" – a naturalist assumes no miracles occur.

tree to fall in the forest and not make a sound). The system acts in accordance to its own volition – whether real or implied. The question I've asked myself is, "What are the biases of the global Technium?" I have come to conclude if we could understand those biases, we could predict what this "collective" intends – what it deems to be its mission and goal.

It logically follows, from my vantage point at least, that since I see this as a life-like system, I find its origins within life itself – life is self-directed to lead to intelligence and perhaps to some level, self-awareness. To say it another way, the Technium is an extension of the same impulse that self-organized into life, and that continues to advance through the Technium, so that the Technium is not *in its essence* anti-life. My perspective (and it may reflect only my view and not those of my fellow authors), is that this collective network of machines, the Technium, derives from and remains compatible with other living things. It may be materially comprised mostly of metal and silicon instead of carbon, but in terms of the Cosmos, it perpetuates, advances, and accelerates the processes that life and evolution (to the extent it operates as a mechanism within the Cosmos) were already undertaking on the planet.

Kevin Kelly says it no less provocatively in this way:

> It's moving in certain directions, and I would say that if we were to make a list of where it's going, it's not a destiny but kind of a direction. **It's moving towards complexity, and it's moving towards more sentience—more mind. It's moving towards more specialization. It's moving towards more energy density, and there's a whole set of other directions that life is also moving towards.** So if we want to imagine where technology will be in 100 years or 1,000 years, we can go down the list and say it's going to be more complex than it is today. There's going to be more minds and artificial minds everywhere. Whatever we make today, we're going to have more specialized versions of it in the future. It becomes more mutualistic, in the sense that technology becomes more dependent on other technologies. **We ourselves, our society, will become more mutualistic. These are all some of the things that I would say technology wants because the system itself is biased**

in these directions, inherently outside of what humans like us want.[13] [Emphasis added]

Kelly has experienced stern criticism for his views on the Technium. But such reaction to his position evinces the breach that traditionally exists between physical and metaphysical realities. His assertions have generated contrary philosophical viewpoints, mostly ardent defenses for the distinction of biological and mechanical forms of life.

Let's return to the example of the collection of neurons creating ideas. Ideas are non-physical constructs. As a Christian, I believe life amounts to more than its material, biological composition. Life involves the spiritual, non-physical elements – namely, a soul – which reveals itself in the physical realm through its space-time manifestation. We can consider an analogy of chalk and a chalkboard. Copious equations and sentences can be written on a chalkboard with a piece of chalk, all communicating ideas on every topic under the sun (and many others beyond our solar center!) Yet, once the eraser sweeps across the chalkboard, the chalk turns to dust swirled into the air and accumulating on the eraser. Molecularly, the chalk particles still exist although the ideas conveyed with these chalk marks vanish just as certainly as the chalk particles scatter. The formation of the chalk markings represents various symbols conveying information expressing letters, numbers, pictures, and equations, but more generally – *ideas*. To communicate, the writer and the reader must share a common interpretation of the symbols independent of whether they're written with chalk on a chalkboard or with a heavy marker on butcher-block paper. Those symbols presented and interpreted represent ideas. Ideas are independent of the chalk and the chalkboard. They transcend our biological hardware (our ocular organs and the grey matter between our ears), but they require the biological "machinery" to send and receive the data.

---

[13] Kelly, Kevin., Edge.org, 'The Technium: A conversation with Kevin Kelly" 2-3-2014 Link: https://www.edge.org/conversation/kevin_kelly-the-technium

In a similar fashion, Kelly asserts a non-physical reality to the network of machines when supposing the collection of mechanical technology evokes or emits ideas that pulsate beyond mere electrical currents and emulates qualities comparable to what we identify characterizes life. And while Kelly's pronouncement may seem absurd when presented before us in the "everydayness" of our typical real-world awareness, there are more and more voices in the scientific community echoing his sentiment. This shift in thinking constitutes a breathtaking redefinition of the nature of existence for future generations, especially as it relates to the message of the Gospel. Whether we applaud it or not, it remains true that just as our understanding of physical matter progresses in the realms of science (whether the large or the small aspects of the Cosmos), so does our appreciation of the meaning lying dormant within the biblical text. In other words, *we believe that what we learn today in physics or astronomy has been in the Bible all along, just awaiting our discovery.* This isn't accidental. The God of the Bible experiences the past, present, and future all as one. He could and did encode truth in the ancient text anticipating that future day we would decrypt it – when our minds were ready and circumstances would lead us to draw such conclusions.

Figure 46 – KEVIN KELLY, MASTER OF THE TECHNIUM

Kelly continues on to affirm a force exists within matter organizing it into *life*. With no explanation offered (let alone evidence of who or what that force might be), Kelly calls upon metaphysical supposition to explain what the pure materialist would regard as nothing more than physical substance without any intrinsic meaning. And while Kelly doesn't propose knowing where the Technium will wind up, he does express faith in its direction, which includes attributes such as intelligence, consciousness, and "energy density" (the amount of energy retained in a specific space expressed in a pertinent the form of units). Indeed, going farther, in the closing portion of the citation Kelly implies that the Technium has a mind of its own.

However, what should concern us comprises much more than the construction of the Technium. Instead, humans should be made aware of our involvement *with it* and response *to it*. We interact with the Technium constantly. We continue to expand its reach and deepen its presence each time we go online. We should not draw the conclusion that the machine in and of itself grows into a monster, as if all machines are evil and the biggest machine is therefore the "baddest" monster of them all. Nevertheless, for Kelly, humans imbue the Technium with what we ourselves constitute in our inner core. Indeed, some that study this phenomenon suggest that the Technium reflects who we are in accordance to our true nature. Frighteningly, we appear to be empowering a life-form mirroring the sinful nature of humanity. So it is, humankind is making a machine in its own corrupted image. As we consider the potential of *transhumanism* with its hoped for integration of biology and information technology, the product of the merger may be most grotesque. What we wind up with may be the very beast of Revelation, *truly a monster of our own making.* A chimera-colossus!

PROPHETIC REFLECTIONS ON THE TECHNIUM BEAST

What Kelly describes as a lifeform bred by the collective effort of human society, also supplies new meaning to the advance of the internet, social media, and "Big Data". Evangelical Bible commentators (past and present) believe God specially created Adam as described in the scripture (that is, he did not "evolve" from prior hominid lifeforms). They suggest Adam was created as an adult human being, complete and equipped with all aspects of a mature *homo sapiens*. Given the ongoing reality of human-designed-and-developed data being constantly uploaded into the ubiquitous aether of the internet (think "the cloud"), a similar threshold may be surpassed at one particular instant, where *self-awareness of the machine* – like the glorification of humanity – begins in the twinkling of an eye (1 Corinthians 15:52).

In the case of the Technium, like the fully grown Adam at the first moment of his creation whose faculties were fully operational, the Technium's faculties will also be fully formed and functioning as an interconnected whole *across the globe*. This may amount to what Kurzweil defined as the technological singularity (also sometimes referred to as the "event horizon" for artificial intelligence). Granted, its mind would exist solely in the collective internet aether. Nonetheless, humanity will interface "face-to- face" with this new digital-yet-sentient lifeform unlike any other created intelligence, i.e., angels, demons, or extraterrestrials from the wildest imaginations of science fiction (or the History Channel!) What is popularly called artificial intelligence (A.I.) will be, from the standpoint of consciousness, artificial life.

It will then exist. At this moment the beast of Revelation arises from the sea and stands on its own two feet (since the Bible uses anthropomorphic qualities to depict the LORD God, we certainly shouldn't hesitate to assign them to the Technium). *It is aware*:

> *And I saw a beast rising out of the* **sea**, *with ten horns and seven heads, with ten diadems on its horns and blasphemous names on its heads.* Revelation 13:1

Commentators suggest that the "sea" here is a symbol for "people of the world" and the chaos generated by their political systems. They also connect it to another entity in Revelation 17 – *Mystery Babylon:*

> *And he saith unto me, The waters which thou sawest, where the whore sitteth, are peoples, and multitudes, and nations, and tongues.* Revelation 17:15

This direct linkage of the sea to the "peoples, multitudes, nations and tongues" proclaims the surprising origin of the beast: *humankind*.

Just to reinforce that the image is used in this way throughout scripture, consider this passage in Isaiah:

> *Woe to the multitude of many people, which make a noise like the noise of the seas; and to the rushing of nations, that make a rushing like the rushing of mighty waters!*

> *The nations shall rush like the rushing of many waters: but God shall rebuke them, and they shall flee far off, and shall be chased as the chaff of the mountains before the wind, and like a rolling thing before the whirlwind.* (Isaiah 17:12-13)

It appears our continued interaction with the information "cloud" has already given birth to the Technium at an embryonic level.

EXAMPLES OF CONFLUENCE – MAN AND MACHINE

When perusing recent news items, I discerned how many scientists, at least at mainstream institutional and academic levels, are moving forward with the push for *the merger of man and machine* without regard for ethical concerns. As a result, many of the stories we encounter today sound more like science fiction than fact. But fact they are, nonetheless.

In the Spring of 2016 on his YouTube channel, this author reported on several unbelievable developments in the world of science and technology.

1. I told my audience of a female "Cyborg Artist" who possesses an implant in her arm that supplies sensations representing seismic activity on Planet Earth. Based on these impulses, she dances to the vibration of earthquakes!
2. NPR (National Public Radio) reported that scientists at UC Davis have been growing human-animal hybrids in the wombs of pigs in an attempt to grow human replacement organs.
3. IBM, certainly one of the most famous computer companies of all time, announced the development of a miracle vaccine that would protect people from the Zika Virus, Swine Flu, Bird Flu, and other feared infectious diseases threatening epidemics among humanity globally.
4. Finally, a private company called *ReAnima* has been given the go-ahead to begin attempting to resurrect, or at least re-animate, clinically dead patients. The occupation of Victor Frankenstein, the original re-animator, has finally sprung to life in the real world.

Once again, what all of these stories have in common is the eerie connection to various mysterious and foreboding passages found in the Bible. With your permission, I will take them up one at a time and respond.

The "Cyborg Artist" goes beyond mechanical or electrical implants to provide "sensing" information beyond sight and sound. Humanity is now merging with machines. The Technium itself will not only occupy digital space (and its "mind" exist in cyberspace), but hardware that can be networked to the "collective" will live in our biological space. The advance of transhumanism has spawned new and unforeseen mechanisms that could become the means to fulfill the Mark of the Beast. This goes far beyond a bar code and an RFID chip. These new interconnections between humans and machine expand our humanity's five senses – creating a sixth, seventh, ad infinitum sense – which could soon extend to the point where humans and machines embrace one another at a level reminiscent of the *Borg* in *Star Trek: The Next Generation*. But is "our resistance futile"? When we are "assimilated" have we become the one with the *Image of the Beast*? Is the Mark of the Beast that minute microchip that connects us to the network and makes us one with everyone else and with "him"? Not only does the Mark of the Beast dictate buying and selling, but privacy would be impossible too. In a sick reflection of the omnipresence of God taught in Psalm 139:7 ("Whither shall I go from thy spirit? or whither shall I flee from thy presence?"), there would be no place we could go where we would escape the awareness of the Beast. That is why stories like the "Cyborg Artist" set off all sorts of alarms for me.

The story from NPR (scientists growing human-animal hybrids) also rings a loud warning bell. They are altering the biological construction of the human being. It is altering our original design. Once again, but in this instance more intentionally, humanity seeks to build a machine in its own image, and the image of man is transforming into the beast.

The method in which IBM has set about the process of creating what may be a universal vaccine parallels one hypothesis regarding how the Mark of the Beast might be deployed. With the combination of computing, synthetic biology, and a required vaccination, we see not only why governments might mandate an injection into the human populations, but the political infrastructure necessary to create a climate of "no buying or selling" or participation in civil society, without the vaccine-cum-Mark of the Beast. Already, we are verging on mandated vaccines in the name of protecting "all the children", not just the ones under any given parent's supervision and care. To be sure, the debate is an important one and valid points can be made on both sides of the argument. But it seems inevitable that, "the needs of the many outweigh the needs of the few" and parental decision making for our children's health will be lost to us.

And finally, regarding the company *ReAnima* attempting to enliven the tissue of clinically dead patients, Is this not the ultimate act of blasphemy towards God? Is this not humanity playing God? Paul warned of the lawless one, the "son of perdition" *"who opposeth and exalteth himself above all that is called God, or that is worshipped; so that he as God sitteth in the temple of God, shewing himself that he is God."* (2 Thessalonians 2:4) This description of the Antichrist reflects his immense hubris. His powers and abilities will attempt to mimic those of the God in heaven. The Bible tells us that God's power was demonstrated through the resurrection of His Son, Jesus Christ (Romans 1:14) as the first of many who will be resurrected. And in the promise of His Second Advent means the resurrection of all the dead in Christ and the glorification of the believer's body. *"But our citizenship is in heaven, and from it we await a Savior, the Lord Jesus Christ, who will transform our lowly body to be like his glorious body, by the power that enables him even to subject all things to himself."* (Philippians 3:20-21)

This glorification is the gift of an immortal body fit for heaven. *"So is it with the resurrection of the dead. What is sown is perishable; what is raised is imperishable."* (1 Corinthians 15:42) ReAnima, probably unwittingly but not innocently, seeks to counterfeit prophetic fulfillment.

CONCLUSION

These are just four such stories. There are countless more. Their recounting demonstrates the world grows stranger than fiction. What was once viewed as mythology or at best fantasies of the future, are now becoming plausible. The eschatological vision of the two beasts described in the Book of Revelation manifests *now in this generation*. The man-machine hybrid has manifested, a chimera of magnificent, let decadent proportions really exits – big as life and twice as ugly!

Perhaps in a way similar to how God "animated" Adam by breathing into his nostrils the "breath of life", the image of the first beast will be animated by the second beast in some manner of breathing into him "life". Perhaps it will be throwing a switch or plugging in a wire. Maybe it will be executing a program developed and made ready just for this purpose. Furthermore, it seems obvious that giving life to the image will be a ceremonial event for all the world to see. We can only conjecture, but our imaginations have been filled with such possibilities thanks to unending bombardment of television programs expressing such visions, special-effect laden cinematic extravaganzas, and memorable prescient stories composed a century ago as I have recounted here.

Contemplating technological methods for fulfilling other prophetic Biblical passages concerning the Abomination of Desolation, the Antichrist himself, and the beast "system" excites deeper investigation to uncover what may soon come to pass. Nevertheless, we can draw a probable conclusion from the one-hundred-year-old Machine of Forster, the Technium of Kelly, and the beast image of Revelation. We head towards a major clash of civilizations, or more likely a confluence between biological life begotten by humans and pseudo-life spawned by electrons and silicon. Could this be the meaning of one of the most sinister passages in Bible prophecy? *"And whereas thou sawest iron mixed with miry clay, they shall mingle themselves with the seed of men: but they shall not cleave one to another, even as iron is not mixed with clay."* (Daniel 2:43)

As we watch the Technium evolve, our attention inevitably shifts to the biggest machine ever built by humanity – the mysteriously strange, nearly infinitely powerful – machine that is the large hadron collider (LHC) in Geneva at CERN and its connection to the previously largest structure ever built by humanity – the Tower of Babel.

We turn our attention next to his these prodigious and yet preeminently *monstrous towers* and why they are connected to today's world, despite a time gap of over 4,000 years.

THE TOWERS OF THE TECHNIUM
Gonzo Shimura

"We must remember this when we are ready to reach for the sky,
God came down and confused man's language and scattered them."

Toba Beta, *My Ancestor Was an Ancient Astronaut*

After the fall of man, Adam and Eve were expelled from the Garden of Eden (Gen. 3:23). It can be theorized that what occultists interpret as the Golden Age of pre-history was the Bible's "time of Eden". Under this premise, these same occultists falsely assert that Yahweh was an evil deity, who kicked Adam and Eve out of paradise in an act of rage and injustice.

The Golden Age was a time when heaven and earth were still one, before Yahweh placed a barrier between the two. The hubris of man has been the pursuit to restore this perfect circumstance of existence; to have heaven and earth together once again. Such a philosophy is evident in the early actions of humanity and its interaction with supernatural beings as recorded in the Bible. After the great flood of Noah, which wiped out the initial race of giants called the Nephilim who were the result of intermarriage between the sons of God and the daughters of men (as discussed by Josh and Doug an in earlier chapter), a tower was constructed under the leadership of the first global world leader, Nimrod. In Genesis, chapter 11, we read of the strange events surrounding the construction of the *Tower of Babel*.

> *And they said to one another, "Come, let us make **bricks**, and burn them thoroughly." And they had brick for stone, and bitumen for mortar.* (Genesis 11:3)

The first aspect of this ancient technology to consider is *the brick*. The word for "bricks" is *leben*, which means brick or tile, but can also mean "tablet."[1] Such a translation can open up to some wild speculations. First, if these were perfectly uniform cut bricks, it begs the question, "What kind of technology these ancients used in order to reproduce mass quantities in such precise geometrically consistent fashion?" Secondly, if "tablet" is the more accurate translation, we can begin to entertain its supernatural and occult significance. For example, there is an *Emerald Tablet* of occult lore. And objects like the "Tablets of Destinies" of Sumerian myth, which supposedly allowed its possessor to become the de facto ruler of the universe, harnessed all power even unto the ability to destroy the Cosmos. While "bricks" from Babel may not have anything to do with such ancient artifacts, the possibility remains that the knowledge which facilitated building the Tower originated from a supernatural source enabling such things.

The *bitumen*, which was used for mortar, also has its significance under the scope of occultism. As the workable paste for building, the mortar was one of the first substances used in building large stone structures. Importantly, it remains an essential element for stonemasons in their most literal definition.

Perhaps some visual examples will help us better understand the Tower of Babel. Let's consider the Great Bath of Mohenjo-Daro, believed to have been built in the Third Millennium BC. Today, the Great Bath sits as a pile of ancient ruins. But many millennia ago, this pit of bricks was employed to collect water and serve as a bath. The main ingredient to make this process possible was the use of bitumen as a lining on the sides and beneath the brick structure.

The word "bitumen" is interchangeable with the similar and familiar substance "asphalt." And while asphalt is made from petroleum extract, the natural substance of bitumen is found on the bottom of lakes

---

[1] GK H4246 | S H3840 & 3843　לְבֵנָה　lᵉḇēnāh n.f. [root of: 4236, 4861]. brick; tablet. "ל," Kohlenberger/Mounce Concise Hebrew-Aramaic Dictionary of the Old Testament, n.p.

and oceans. Attaining the substance in its natural state would be tremendously difficult even for advanced civilizations. Evolutionary minded contemporary scholarship suggests Neanderthals first used it some 40,000 years ago. It would have us believe bitumen became a staple substance as an adhesive and waterproofing material for the next 30-plus millennia, leading up to its widespread use in the ancient Near East. Such naturalist notions notwithstanding, we believe it was used in *building the Tower of Babel*. In fact, the use of bitumen goes back to Noah and his building the Ark. Logically, its usage may go back to the days of Cain, the son of Adam, credited as the first city builder. So where did Cain obtain the know-how to acquire and use bitumen as an adhesive substance? An account from the pseudepigraphical *Book of Enoch* possibly sheds light on this issue:

> *And restore the Earth which the Angels have ruined. And announce the restoration of the Earth. For I shall restore the Earth so that not all the sons of men shall be destroyed because of the knowledge which the Watchers made known and taught to their sons.* (Enoch 10:7)

Figure 47 - THE BRICK WORK OF THE GREAT BATH

A portion of this information may have accentuated the hubris of mankind when laying the bricks for the Tower of Babel – a tower that could to reach into heaven – whose meaning can't be taken literally (many critics presume all conservative biblical scholars think it so).

In the image above, notice how the brickwork of the Great Bath[2] appears as refined as brickwork today. Recall this was 5,000 years ago (as dated by mainstream archeologists). Whether before or contemporary to the building of the Tower of Babel, it indicates the level of complexity, planning, and purpose that went into such ancient projects and was implicit in the building of what I would identify as the "first *Technium*". It also recalibrates our understanding of the level of sophistication present among humanity during ancient times.

THE JASHER TECHNOLOGY

The *Book of Jasher* records details about the Tower of Babel that are both informative and stunning especially when one considers the implications of a civilization that may have been modest technologically given what we know about their society as a whole. And yet, it may have been very highly advanced in one specific area, accessing some sort of *cosmic scalar technology* – an exotic and unexpected weapon of mass destruction – possibly technology passed down from the antediluvian world.[3] No matter what shape such a technology might

---

[2] The **Great Bath** is one of the best-known structures among the ruins of the ancient Indus Valley Civilization at Mohenjo-Daro in Sindh, Pakistan. Archaeological evidence indicates that the Great Bath was built in the 3rd Millennium BC, soon after the raising of the "citadel" mound on which it is located. See https://en.wikipedia.org/wiki/Great_Bath,_Mohenjo-daro.

[3] "**Scalar waves** were originally detected by a Scottish mathematical genius called James Clerk Maxwell (1831-1879) He linked electricity and magnetism laying the foundation for modern physics, but unfortunately the very fine scalar waves (which he included in his research) were deliberately left out of his work by the 3 men, including Heinrich Hertz, who laid down the laws taught for physics as a discipline at colleges. They dismissed Maxwell's scalar waves or potentials as 'mystical' because they were physically unmanifest (sic) and only existed in the 'ethers' and so were determined to be too ineffectual for further study. These enigmatic... scalar waves may have been forgotten except that Nicola Tesla accidentally rediscovered them. Tesla later experimented using the research of the German Heinrich Hertz, who was proving the existence of electromagnetic waves... Tesla found, while experimenting with violently abrupt direct current electrical charges, that a new form

have taken on in the days leading up to the Great Flood, the point I wish to make here is that we are revisiting the *days of Noah* once more. Such technology inevitably defines human potential while at the same time portending a new age, no doubt to be seen by some as an advancement of our species despite the fact this technology might be used only as a more efficient means to destroy life on this planet.

In chapter 9, verse 21 of *Jasher* it states:

> *And all the princes of Nimrod and his great men took counsel together; Phut, Mitzraim, Cush and Canaan with their families, and they said to each other, "Come let us build ourselves a city and in it a strong tower,* ***and its top reaching heaven, and we will make ourselves famed, so that we may reign upon the whole world, in order that the evil of our enemies may cease from us, that we may reign mightily over them***, *and that we may not become scattered over the earth on account of their wars."*[4]

Figure 48 - A PAINTING BY PIETER BRUEGEL THE ELDER CALLED "THE TOWER OF BABEL (VIENNA)."

---

of energy (scalar) came through." See http://www.tokenrock.com/explain-scalar-wave-technology-77.html.

[4] Retrieved from http://www.succatyeshua.nl/upload/files/The%20book%20of%20Jasher.pdf (emphasis mine)

We are tempted here to speculate that what we are encountering may literally be *wormhole technology* – a means to traverse the heavens, aka outer space – or to enter other dimensions (which likely overlap). They suppose this project enhances their reputation (they see it improving their status one supposes among the demigods of either the Nephilim or the Divine Council as discussed previously. It vexes Nimrod to know that his "enemies" and their "wars" could prevent humankind from achieving his end, indicating he understood that the kingdom of Yahweh was ultimately more powerful still – for Yahweh remained the Most High God. Perhaps Nimrod thought he could "slip one past Him."

While popular images of the Tower of Babel comprise merely a tall building (the famed Ziggurat), the Bible hints the technology on the plain of Shinar was much more complex:

> *And the LORD said, "Behold, they are one people, and they have all one language, and this is only the beginning of what they will do.* ***And nothing that they propose to do will now be impossible for them.*** (Genesis 11:6, emphasis added)

Again we must ask, "What manner of technology would cause God to state that humanity would be able to do whatever it proposes to do?" Here the *Book of Jasher*, verse 25, supplies a helpful clue:

> *And the building of the tower was unto them a transgression and a sin, and they began to build it, and **whilst they were building against the LORD God of heaven, they imagined in their hearts to war against him and to ascend into heaven.**[5]*

This passage reminds us of what we read in Isaiah's famous passage in Isaiah, chapter 14 in which Satan asserts his infamous five "I wills" – the most applicable of the five being: "You *said in your heart, 'I will ascend to heaven; above the stars of God I will set my throne on high; I will sit on the mount of assembly in the far reaches of the north..."*[6]

---

[5] [3] Ibid. [Emphasis Mine]

[6] [4] Isaiah 14:13

Many have dismissed such verbiage as nothing more than allegorical boasting, negating the obvious possibility that these passages could inform us that Satan in the ancient past, or the children of men in the time of Babel, sought to ascend into heavenly dimensions, to launch some sort of attack upon God. This incredible notion returns us once more to contemporary events. We live in such days. We scan the horizon half-expecting to witness a braggadocios beast arise from the chaotic sea as depicted in Revelation 13.

In a lecture entitled "UFOs, the Tower of Babel Moment, and Space Collateralization and Commercialization", Dr. Joseph P. Farrell at the *Secret Space Program Conference* (2014), employed the Tower of Babel as a metaphor to suppose that human leaders, like Nimrod, have recently sought to emulate exotic weapons "from extraterrestrials" as a means to develop a defense against these ETs that want to do more than just "phone home" – Farrell supposes they threaten our very existence:

> Anytime you have a technology that is capable of engineering the fabric of space time, or the zero-point energy, locally, on the laboratory bench, you have first of all a technology that if weaponized would make the hydrogen bomb look like a firecracker. If you engineer it for propulsion purposes, then quite literally the sky is the limit. And thirdly…this would mean a complete restructuring of the financial system as necessary…they [a powerful elite who possesses such technologies] are in the process of erecting a new system. So the question is, "Why?" And I would submit…that the reason is, the UFO is the great hole in our contemporary culture and particularly in our understanding of technological developments…this would have led to the realization…that humanity was in a danger period; a Tower of Babel moment of human history, during which non-terrestrial [a force not of this world] intervention prior to such human technological emulation in human affairs and independence, had to be contemplated and prevented…[7]

Farrell goes on to suggest that nowhere in Genesis 11, in the account of the Tower of Babel, does it mention the building of the Tower was in any way *sinful*. But as we know, while the *Book of Jasher* is not consid-

---

[7] Joseph P. Farrell, UFOs, the Tower of Babel Moment, and Space Collateralization and Commercialization., Secret Space Program Conference, San Mateo CA June 29, 2014.

ered inspired text, it does profess the axiom of monotheism and a supreme God, Yahweh, as well as contending that the project's intent could be nothing other than a sin against Him. *"And the building of the tower was unto them a transgression and a sin."* What we should gather from the testimony of the *Book of Jasher* accompanied by the passage from Isaiah, chapter 14, is that *the activity of Satan and the spirit of antichrist are in play in these matters.* As a result, while I disagree with Farrell about whether or not building the tower was a sin; I would endorse his observation that technologies fell into the laps of the builders of the Tower of Babel from ancient ancestors ... and such technologies to oppose the God of Gods are under development today! The so-called *breakaway civilization*[8] more likely than not possesses them already. And taking it one step further, the capabilities of these technologies – if humankind were to continue to exist for an indefinite period – will inexorably launch us into the stars. In fact, many believe it is *a fait accompli* – that the breakaway civilization already traverses the Cosmos, perhaps employing the same cosmic scalar technologies that the ancients once did. Indeed, although beyond the scope of this chapter, evidence exists to suggest a secret program for space colonization and weaponization sits under our noses (perhaps more aptly depicted as "over our heads!"). The question is, "How advanced are these technologies?" This issue is something to tackle another time. But I regard it real enough to challenge the reader to consider the prophetic implications with me here.

AN ASSAULT UPON YAHWEH & HIS HEAVENLY HOST

The *Book of Jasher,* chapter 9:29 states:

> *And the LORD knew their thoughts, and it came to pass when they were building they cast the arrows toward the heavens, and all the arrows*

---

[8] The notion of a "Breakaway Civilization" was coined by Richard Dolan after his exhaustive and extensive research into (sic) his books **UFOs and the National Security State Volumes 1 and 2.** He came to the realization that the deep secretive black budget world had actually become a civilization on its own with its own exotic technology, a different view of the cosmos, our place in it, and a different version of human history. This is a new avenue of research that... delve(s) into the Occult, secret societies, central banking, UFO secrecy, the Military Industrial Complex, and the Secret Space Program comprising this Breakaway Civilization." See http://breakawaycivilization.com/2012/08/15/what-is-the-breakaway-civilization/.

fell upon them filled with blood, and when they saw them they said to each other, "Surely we have slain all those that are in heaven." [9]

What kind of arrows were aimed and launched towards heaven? And who were they trying to destroy? If we understand this passage correctly, Nimrod and his followers fired some type of projectile into the heavens, in hopes to "bring down" Yahweh and His heavenly host. It seems odd that those on the ground lost sight of the arrows, suggesting that this projectile might have been larger in scale than one might conjecture, OR it passed beyond the skies and into a portal or *star gate*. And notice how their weapon returned to earth "filled with blood." This reference might be allegorical, but if literal, then something was done to the projectiles to make them appear as if they had hit their target. After Nimrod saw that the arrows were "filled with blood," they assumed that the enemy was maimed or eliminated.

Was this blood *literal* blood? Perhaps the blood was from the children of men when "all the arrows fell upon them," as they eventually caused more harm to themselves than taking out the LORD and His Host. It can also be argued that this is a well-disguised extra-biblical example of the Evangelion – a theological assertion regarding sacrificial love through the shed blood of Jesus Christ. Then again, perhaps the most extraordinary elements of the story should be dismissed as they stand too far outside any reasonable consideration.

Secular speculation, particularly those who hold anti-Christian presuppositions, might consider "those that are in heaven" are *merely* (as if merely could be the right word!) extraterrestrials with highly advanced technology, such that they could know "our thoughts" and either prevent, or assist in developing advanced technologies. But if this event was in fact an ancient shadow of things to come soon upon the earth, the implications would be, obviously, most significant.

---

[9] Http://www.succatyeshua.nl/upload/files/The%20book%20of%20Jasher.pdf (emphasis mine)

Farrell's thesis essentially is this: the elite "underworld" has already moved beyond a contemporary equivalent of the biblical and historical "Tower of Babel" event. That is, they have arrived at a point in human history where humanity's possibilities appear to be without limits, most likely as a result of techno-scientific advancements. Some researchers would have us believe it would be in the best interest of humanity, if we prohibit intervention from a God such as Yahweh with His moral requirements. We should not allow Him to prevent men and women from achieving our "ends" this time around.

Assuming this motivation lies behind the conscious choice of this presumed elite, they would have then effectually adopted Satan's mission, as outlined in Isaiah 14, anticipating the God of heaven cannot keep them from seeking that to which they have pledged themselves. They persist in hostility to Yahweh and to Jesus His Son. Therefore, in terms of the biblical report, it should not be surprising that they represent what the Bible defines as "the spirit of Antichrist". When humans embark on an arcane path like this, a paradigm shift occurs. It becomes a nefarious commitment, obvious to Yahweh's followers, that these actors will worship the Old Dragon, just as the Bible foretells.

HEAVENLY HOSTS AND THE BABEL CONNECTION

In Genesis chapter 11, we saw that God interrupted the building of the Tower of Babel. Similarly, from the *Book of Jasher* we garner clues to what actually transpired. Perhaps as punishment as well as prevention, the demigods Nimrod feared acted in accordance to the directives of Yahweh to confound the attempt of humankind to ban together – establishing themselves in an exalted state upon the Earth:

> *And God said to the **seventy angels** who stood foremost before him, to those who were near to him, saying, "Come let us descend and confuse their tongues, that one man shall not understand the language of his neighbor, and they did so unto them."* [10]

---

[10] *Book of Jasher.* Chapter 9, Verse 32 http://www.succatyeshua.nl/upload/files/The%20book%20of%20Jasher.pdf

The mention of *70 angels* at this juncture is most significant because the number just happens to equate to the number of nations God identified in the so-called Table of Nations contained in the *Book of Genesis* chapter 10. As we discussed in an earlier chapter, Dr. Michael Heiser argues that these 70 composed the **Divine Council,** a heavenly bureaucracy. This Council eventually came to oppose Yahweh, seeking to be worshipped by men as gods. This would ultimately lead Yahweh to choose Israel as His own *peculiar* nation through which His promises to redeem humanity would be fulfilled. Heiser states:

> The OT exhibits a three-tiered council... In Israelite religion, Yahweh, at the top tier, was the supreme authority over the divine council, which included a second tier of lesser *elohim* ("gods"), also called the "sons of God" or "sons of the Most High." The third-tier comprised the *mal'akhim* ("angels")."[11]

Figure 49 - VIRGIL SOLIS: "GOD'S COUNCIL"

There are many proof texts in the Bible indicating these tiers exist within the Council and that there was no shortage of drama leading to Yahweh separating Himself from the others, choosing Israel as His own nation. We see the most notable occasion within Psalm 82:

> *¹ God has taken his place in the **divine council**; in the midst of the gods he holds judgment:*

---

[11] Heiser, Michael S., *Old Testament Godhood Language,* The Divine Council

² *"How long will you judge unjustly and show partiality to the wicked? Selah*

³ *Give justice to the weak and the fatherless; maintain the right of the afflicted and the destitute.*

⁴ *Rescue the weak and the needy; deliver them from the hand of the wicked."*

⁵ *They have neither knowledge nor understanding, they walk about in darkness; all the foundations of the earth are shaken.*

⁶ *I said,* **"You are gods, sons of the Most High, all of you;**

⁷ **Nevertheless, like men you shall die, and fall like any prince."**

⁸ *Arise, O God, judge the earth; for you shall inherit all the nations!*[12]

Here we clearly see Yahweh addressing His heavenly council in judgment. The gods appointed to oversee the 70 nations seemed to have failed at the work they were assigned. As a result, Yahweh pronounces that they will die like men and "fall like any prince."

You may be wondering how we know that the gods Yahweh addressed in this Psalm and elsewhere are the 70 to whom authority over the nations had been given. Deuteronomy 32:8 tells us that this was in fact so:

> *When the Most High gave to the nations their inheritance, when he divided mankind, he fixed the borders of the peoples according to the number of the sons of God.*[13]

This brings us full circle back to the time of Babel, because here we see the *Most High* giving each nation their inheritance: dividing the people and giving them fixed borders according to the number of the sons of God – these 70 elohim described in the *Book of Jasher*. Could this event and the confusion of the languages at the Tower of Babel, be one and the same? If so, then the assignment of people groups to the servants of Yahweh – His "host" – was meant to be a means to bring order to humankind after the chaos ensuing from the confusion of languages. However, it seems even the hosts of God were subject to temptation, perhaps falling into the same sin which occurred in the days of Genesis 6 (this might be

---

[12] Psalm 82

[13] Deuteronomy 32:8 ESV

yet another explanation for the "second incursion" – how the Nephilim appeared once more after the flood).

Dr. Heiser distinguishes between the "second-tier" Divine Council from the "third-tier angels". It's remarkable that virtually all other scholars have NOT pointed out this distinct *class of beings in the Old Testament.* This failure leads to identifying the Divine Council members as mere angelic beings. The New Testament provides a hierarchical structure absent in the Old Testament (i.e., "powers and principalities"). We know that there are legions of angels[14] – that there are many more than 70. So it would seem a gross mistake to confuse these distinct classes of beings.

Recall also that the *Book of Job* describes the sons of God rejoicing at the moment when the earth was created (Job 38:7) indicating that they existed before the Earth's creation and its human caretakers. Exactly when the Divine Council was put into place isn't spelled out for us. But its finite number suggests that there was a distinction made between angels and the "Council members".

We ought not be alarmed entities of higher intelligence exist and interact with us today. And while there are many forms of spirits, ghosts, and a plethora of paranormal accounts, testimony, and first-hand stories, we must remember that the very instruction manual for life, the Bible, speaks of such fringe and unfamiliar experiences with explicit explanatory power. It reminds us despite all of the evils that have befallen humanity's dark history, God always has a plan, and it is in play today.

Moving on: it's intriguing to speculate on "what transpired" and when the Tower of Babel was built by its ancient agents. However, it's more critical to note parallel events repeating themselves right before our eyes.

INTO MODERNITY

The Book of Jasher also records the destruction of the tower itself, soon after the dispersion of the people:

---

[14] *"Do you think that I cannot appeal to my Father, and he will at once send me more than twelve legions of angels?"* (Matthew 26:53) Assuming of course that Jesus reflected the thinking of the Jews of his day, or in fact, he could literally accomplish this.

> *And as to the tower which the sons of men built, the earth opened its mouth and swallowed up one third part thereof, and a fire also descended from heaven and burned another third, **and the other third is left to this day, and it is of that part which was aloft, and its circumference is three days' walk.*** [15]

Here we see God bury a third of the tower, burn a third of the tower by reigning fire down from heaven upon it (potentially mimicked by space technology the False Prophet will use as deceptive signs and wonders in Revelation 13), and finally leaving a third of the Tower "to this day" memorializing the event (Note: we don't know exactly when the *Book of Jasher* was composed. We may presume it was written about 1,000 years after the "Babel event" and perhaps 300-500 years after Moses wrote his brief account provided in Genesis). Jasher describes the size of the remaining third of the tower, which is said to be a circumference of a three-day walk. Whether time inflated the size of the Tower to these gigantic, indeed "mythological proportions" we can't say.

According to Talmudic traditions, a day's journey of the average man was called *derekh yom,* which equals to approximately 10 *parasangs*. A single *parasa* is considered to be about 2.4-2.88 miles.[16] This means in a *derekh yom*, one could walk somewhere between 24 to 28.8 miles. In three *derekh yoms*, this would add up to somewhere between 72 and 86.4 miles. This is very close to what modern people assert is the distance they can walk over three days' time.[17] Now this would indicate something far different from the traditional views of the Tower of Babel being little more than a tall Ziggurat-like "skyscraper". Rather, it would seem that whatever was being built required unfathomable manpower.

---

[15] *Book of Jasher.* Chapter 9, Verse 38, See http://www.succatyeshua.nl/upload/files/The%20book%20of%20Jasher.pdf.

[16] Wiki: *Biblical and Talmudic units of measurement,* http://en.wikipedia.org/wiki/Biblical_and_Talmudic_units_of_measurement.

[17] When asked how far humans can walk in a day, 25 miles to 30 miles was a common answer with examples given. We could assume a 80-90-mile circumference would not be out of the question with about 26-28 miles across as the diameter. This would mean the base may have been 2 to 3 times that size. It would have been a really horrifically big tower at the base – if mythological "inflation" did not occur in Jasher's account. See https://uk.answers.yahoo.com/question/index?qid=20080110102650AAxYDZF.

And as our investigation moves from these ancient glimpses of the distant past to modern day developments, we can begin to see the type of technologies that must have been harnessed by the ancients which ultimately caused God to terminate their most gargantuan development.

PARTICLE ACCELERATORS

Back in 1952, just seven years after World War II and early in the development of the *breakaway civilization,* the Conseil Européen pour la Recherche Nucléaire, or more commonly referred to as CERN (Council for Nuclear Research) was formed. The mandate of the group was to establish a world-class fundamental physics research organization in Europe. CERN has come a long way since the idea to formulate such a group was proposed back in 1949 when the focus was purely on understanding the inside of the atom (hence the inclusion of *nuclear* in its name). As Anthony discussed earlier, today its main focus has been on particle physics, which includes looking for antimatter, and searching for the Higgs Boson, more commonly known as the "God Particle." In order to conduct this research, its main tool has been the LHC.

Figure 50 - THE RINGS OF ACCELERATORS AS INFINITY (8)

ADVANCED CIVILIZATIONS

The size and other attributes of this world-renowned particle accelerator are impressive:

> The LHC consists of a **27-kilometre ring** of superconducting magnets with a number of accelerating structures to boost the energy of the particles along the way. Inside the accelerator, two high-energy particle beams travel at close to the speed of light before they are made to collide. The beams travel in opposite directions in separate beam pipes – two tubes kept at ultrahigh vacuum. They are guided around the accelerator ring by a strong magnetic field maintained by superconducting electromagnets. The electromagnets are built from coils of special electric cable that operates in a superconducting state, efficiently conducting electricity without resistance or loss of energy. **This requires chilling the magnets to -271.3°C – a temperature (virtually) as cold as outer space.** For this reason, much of the accelerator is connected to a distribution system of liquid helium, which cools the magnets, as well as to other supply services.[18]

It is interesting that the designers of CERN would create conditions as cold as space itself;[19] however, the more relevant issue for our purposes relates to the size of this accelerator. With a circumference of 27 kilometers, or 16.777 miles, this is a magnificently large construction for simply colliding particle beams. Furthermore, aside from trying to recreate the conditions of the Big Bang, the folks at CERN are quite literally trying to rip the fabric of space-time and look through it into other dimensions. They say that "curiosity killed the cat" (while satisfaction brought him back). But just what are the implications for the cosmically curious outside Geneva?

---

[18] http://home.web.cern.ch/topics/large-hadron-collider

[19] There are those who claim temperatures below absolute zero, -273 degrees C. The founder of D-Wave Quantum Computers claims his computer is 150 degrees below the temperature of outer space. This is scientifically impossible according to reliable sources. "What is normal to most people in winter has so far been impossible in physics: a minus temperature. On the Celsius scale minus temperatures are only surprising in summer. On the absolute temperature scale, which is used by physicists and is also called the Kelvin scale, it is not possible to go below zero —at least not in the sense of getting colder than zero kelvin. According to the physical meaning of temperature, the temperature of a gas is determined by the chaotic movement of its particles —the colder the gas, the slower the particles. At zero kelvin (-273 C) the particles stop moving and all disorder disappears. Thus, nothing can be colder than absolute zero on the Kelvin scale." See http://www.rdmag.com/news/2013/01/temperature-below-absolute-zero.

In an article published on January 1, 2013 by *The Guardian* entitled, "Higgs Boson Was Just a Start for CERN'S Atom Smasher – Other Mysteries Await", Andy Parker, Professor of High Energy Physics at Cambridge contends:

> "What you'd expect is that as you reach the right energy, you suddenly **see inside the extra dimensions,** and gravity becomes big and strong instead of feeble and weak...There could be a whole universe full of galaxies and stars and civilizations and newspapers that we didn't know about...That would be a big deal."[20]

It certainly would be a big deal. But perhaps this entire organization is simply a front to develop the kinds of particle laser beams that will be necessary to combat a future invasion from space – or a different dimension entirely. Speculating about such things seems not so crazy after reading other comments made by the scientists working at CERN.

In an article published by *The Telegraph* in 2008 entitled, "Time Travelers from The Future Could Be Here in Weeks," coinciding with the launch of the LHC, many comments were made about the potential of this device to create wormholes. One such example:

> Prof Aref'eva and Dr. Volovich believe the LHC could create wormholes and so allow a form of time travel. "We realized that closed time like curves and wormholes could also be a result of collisions of particles," Prof Aref'eva says.[21]

Then there was mention from the website ScienceGuardian.com, in an article entitled "Large Hadron Collider Challenges the Gods", that there are apparently...

> Three dangers inherent in the CERN insistence on going full speed ahead despite all the warning signals: *strangelets* turning the planet into a smoking asteroid the size of a baseball park, a micro black hole swallowing the earth from the core outwards, if not the sun, and/or the

---

[20] *The Guardian*, "Higgs Boson Was Just a Start for CERN's Atom Smasher – Other Mysteries Await", http://www.theguardian.com/science/2013/jan/01/higgs-boson-large-hadron-collider.

[21] *The Telegraph:* "Time Travelers From The Future 'Could Be Here In Weeks" http://www.telegraph.co.uk/science/large-hadron-collider/3324491/Time-travellers-from-the-future-could-be-here-in-weeks.html.

ADVANCED CIVILIZATIONS

generation of a huge amount of energy equivalent to a thermonuclear bomb per second. [22]

Encouraging words from these leading scientists, aren't they? But there's more. Like many other researchers, we don't believe that it is accidental that sitting in front of the CERN headquarters is a statue of the Hindu god Shiva, also known as "the Destroyer" or "the Transformer" or that the "mother of all machines" sits atop the ancient Roman temple to *Apollyon/Abaddon* – also known as "the destroyer". The Shiva stature seems to comprise **a mission statement in stone,** if you will. Also noted: the suspicious display of ancient text surrounded by a blue laser beam in the foyer of the facility. Consequently, we must ask, "If a 17-mile ring has the potential for such catastrophic destruction, what would it mean for a collider almost triple the size?"

DESERTRON

The **Superconducting Super Collider** (SSC) otherwise known as the *Desertron*, was in the works to be built in the vicinity of Waxahachie,

Figure 51 - THE PARTICLE ACCELERATOR THAT NEVER WAS.

---

[22] See http://www.scienceguardian.com/blog/large-hadron-collider-challenges-the-gods.htm

Texas back in the late 1980s. This project was working towards producing nearly double the amount of energy that the LHC possesses in its incarnation prior to October 2015. In order to do this, the planned ring was estimated to be a circumference of 87.1 kilometers, or 54.1 miles! That's almost three times the size of the current LHC.[23]

Being such a public affair, however, funding for the Desertron was cancelled in 1993 by Congress. While there no direct way to prove it, it is probable that projects like Desertron went deep underground, both literally and figuratively, funded heavily by the "breakaway civilization", in order to reconstruct the capabilities that might have existed through the structure of the ancient Tower of Babel.

If the *Book of Jasher* truly gives us clues about how large the Tower of Babel was, then perhaps it was some kind of exotic machine, perhaps even an ancient particle accelerator using technology of "the gods". We would assume such speculation is ridiculous because the

Figure 52 - THE LARGEST STONE AT BAABEK IN LEBANON
Photo by Ralph Ellis

---

[23] http://en.wikipedia.org/wiki/Superconducting_Super_Collider

amount of electrical power needed defies imagination. But who says that electrical power, i.e., mustering the power of the aether, wasn't something the ancients knew how to accomplish? If the numbers were anywhere near accurate, then the ring making up the Tower of Babel would have been of a circumference that is literally, *over four times the size of the proposed Desertron.*

While this is breathtaking, it is not out of the realm of possibility given what the ancients seem to have been capable of doing, seemingly creating structures that surpass our ability to achieve today (e.g., The Giza Pyramids in Egypt, the Temple of Baalbek in Beqaa Valley, Lebanon). And if such technologies were real in the past, it is no surprise that those who know about them would want to reconstruct these technologies today, in order to begin ripping space-time apart – peering into other dimensions – or worse yet, creating a weapon having the power to fire projectiles into heaven and attempt to, if it were possible, kill God just as Nimrod and his princes attempted (according to the *Book of Jasher*). When the Messiah returns to fight the Antichrist at the Battle of Armageddon, this may be the weapon assembled to greet Messiah and His mighty ones.

BRIDGES TO META-REALITY: CONNECTING THE DOTS

We will now examine the precipice, the Event Horizon, the bridge that connects the furthest reaches of the universe to the smallest particles visible. The topic to bring us this point is the *black hole*. Black holes bend the philosophy of materialism. In fact, it might break it altogether. According to mainstream science, black holes are likely a gateway into other dimensions. Stephen Hawkings recently made an appearance at Harvard University. Commenting on the current state of understanding the Cosmos, in particular the *black hole*, the words of Hawkings subtly prepare us for the many unimaginable prophecies found in the Bible.

> "It is said that fact is sometimes stranger than fiction, and nowhere is that more true than in the case of black holes…Black holes are stranger than anything dreamed up by science fiction writers, but they are clearly

matters of science fact… For more than 200 years, we have believed in the science of determinism, that is that the laws of science determine the evolution of the universe… If information were lost in black holes, we wouldn't be able to predict the future because the black hole could emit any collection of particles.

"It might seem that it wouldn't matter very much if we couldn't predict what comes out of black holes — there aren't any black holes near us…But it's a matter of principle. If determinism — the predictability of the universe — breaks down in black holes, it could break down in other situations. Even worse, if determinism breaks down, we can't be sure of our past history either. The history books and our memories could just be illusions. It is the past that tells us who we are. Without it, we lose our identity." [24]

The suggestion that black holes are mere science fact, is a bit presumptuous; nevertheless, it is true that fellow humans are collecting data, attempting to explain it, and eventually, reconstruct its implications. But the deeper element of what Hawking said concerns the destruction of the past and revising reality with history reconstructed.

There are two things that can be surmised from this quote:

1. The realities being discovered by modern science are changing our fundamental understanding of the Cosmos. Reality is being revised.
2. This monumental shift will also change how humankind defines itself.

First, it's important to point out that by calling *black holes*, "stranger than anything dreamed up by science fiction writers," for all intents and purposes Hawking *confirms the supernatural and metaphysical properties the creation possesses.* Thus, a logical syllogism pops up:

1. Atheists and materialists accuse believers in the Bible's supernatural elements of believing myths and fairytales. However,
2. Science fiction also consists in tales transcending reality; and yet reality surpasses some of the most imaginative notions in science fiction.
3. Therefore, as Science comes to understand the creation correctly, it confronts the same kind of realities expressed in the Bible and other ancient text. This overturns the accusations made against the "supernaturalist".

---

[24] http://news.harvard.edu/gazette/story/2016/04/hawking-at-harvard/

In other words, Science must loosen its moorings, forsake its standard "demythologizing" bias, be liberated from its old orthodoxy, and seek new answers informed by the plethora of supernatural events recorded in ancient texts – events it may then deem ring true. All individuals, including scientists, should be freed to seek out new vistas of the Cosmos. Make no mistake: the "creation paradigm" proposed in this book stands firmly rooted in the Word of God. It is a viable alternative in the marketplace of ideas too. It not only provides a foundation for comprehending the universe, it also makes provisos for reconciling current conundrums or contradictions owing to its present state of corruption.

In contrast, other scientists have explored realms of inquiry and integrated paranormal premises these into their philosophical disposition. From arcane affirmations derived from the mystical practice of alchemy – esoteric concepts promulgated in secret societies – to self-serving New Age perceptions, we may witness a merging of non-biblical supernatural paradigms into a cosmology justified by scientific endeavor which still conflicts with the Christian worldview proposed in this volume. As biblical prophecy predicts an end-times' one-world religious system, it is no stretch to suggest we have encountered the genesis of a demonically inspired deception within present-day science shepherding a flock of self-assured and unsuspecting "sheeple".

If this deception holds serve, then in the coming years there will be a massive accumulation of souls made ready for hell with a most dismal crescendo blaring loudly when that last trumpet sounds on the Day of the LORD.

WHAT'S A SOUL TO DO IF BLACK HOLES HOUND US?

It is at this point we transition from the "meta-outlook" of the Cosmos to the "micro-inlook" (to coin a phrase), concerning the individual self, the place where our identity resides. The ripple effect from the revised "anti-Christian cosmological argument" eventually impacts the realm of personal meaning. It will summarily change how humanity defines itself and how the mass of humankind chooses to think of itself.

Hawkings theorizes our philosophical breaking point will occur when we begin to internalize the knowledge of black holes – in other words, when the reality of what they imply "sinks in". He describes it as the logical inevitable conclusion of determinism:

> Determinism, in philosophy, theory that all events, including moral choices, are completely determined by previously existing causes. Determinism is usually understood to preclude *free will* because it entails that humans cannot act otherwise than they do. The theory holds that the universe is utterly rational because complete knowledge of any given situation assures that unerring knowledge of its future is also possible. [25]

However, my commitment to "freedom and dignity" is undaunted by his cosmic pessimism. I believe the reality remains that *free will exists in the mist of the extremes of determinism.* That is, logically not all choices are available to us – they never are. We are finite and our choices will always be limited. And yet much ability to choose remains at our disposal. We can make unprogrammed decisions. The circumstances in which we find ourselves, where we are in any given situation – whether seen physically or morally – are a result of previously existing choices we have made (as well as many other causes that are not within our control). However, it does not completely negate free will at the present moment as we encounter it. While the circumstances that lead to a set number of choices are more or less fixed, our individual decision making remains our responsibility. Therefore, *free will remains intact.* In contrast, what Hawkings suggests is that black holes devastate our cherished concepts concerning the Cosmos – both scientific and religious – as worked out by humanity's greatest minds across the span of our existence. This attack on what we would count as the *rationality* of existence by the Overload of Materialism and the cancer of black holes causes an identity crisis.

Many perspectives on the outlook for humanity, some of them mentioned earlier, generally share one common thread: seeking perfection. The idea that mankind will someday achieve a higher state of

---

[25] http://www.britannica.com/topic/determinism.

being comprises the driving force for virtually all metaphysical philosophies other than nihilism. In fact, although uncomfortable for some and incomprehensible to others, the Bible promises humankind will be restored one day to a paradisiac preeminence, above the angels, likely above the "council" too, replacing them and seated next to God Himself. The main difference between the Gospel's solution and the solution proffered by occult and mystical philosophies regards *the method by which that higher state is realized.*

The phrase "works based salvation" sounds familiar to churchgoers. It is a descriptive phrase serviceable to express the difference between the Gospel's message of salvation compared to how other religions state their means for "deliverance" from the problem at the heart of humanity's "alienation" from God, from others, and from ourselves. The born-again Christian believes that his or her personal sin was placed upon the shoulders of Jesus Christ. His death was vicarious – it was for us. When He was beaten with whips and crucified on the cross – a transaction occurred that for most of us seems an impenetrable mystery – God imputed (assigned) our sin to Jesus. He died in our place. And on the third day, God demonstrated through the bodily resurrection of Jesus, that this transaction worked just as God planned. Our sins were forgiven. The justice of God was placated. Just as Christ was raised from the dead signifying that the work He had to do on our behalf was completed (the debt was paid in full), we too will be raised from death signifying that we are not accountable for our sins any longer. Likewise, as Christ was destined for heavenly glory, we too are promised that same destination. This event demonstrated a power that is unequaled, even surpassing the unfathomable power of the black hole. Death has been overcome! The Bible proclaims that this very power holds sway over nature, over creation, and over death. God demonstrated in the resurrection of Jesus, that Jesus lives and He lives in us!

> [9] *But ye are not in the flesh, but in the Spirit, if so be that the Spirit of God dwell in you. Now if any man have not the Spirit of Christ, he is none of his.*
>
> [10] *And if Christ be in you, the body is dead because of sin; but the Spirit is life because of righteousness.*

> *[11] But if the Spirit of him that raised up Jesus from the dead dwell in you, he that raised up Christ from the dead shall also quicken your mortal bodies by his Spirit that dwelleth in you.* (Romans 8:9-11)

Our identities now are rooted in Jesus Christ. This power described in Romans, chapter 8, resides beyond human understanding, which means it exists beyond creation, beyond the reality of time, and exists outside the confines of determinism.

Hawkings, in introducing the breakdown of information, posits a transcendent answer – one that cannot be provided by any methods of man's mystical, occult, philosophical, scientific, or technological means. This brings us back to the discussion of technology. Specifically concerning virtual realities.

The virtual realities we have constructed are very much illusory. With the advent of virtual technologies to access these created spaces through our senses, we have already altered how we experience reality. This massive social movement towards the access of these technologies opens up collective gateways into virtual spaces. In this pseudo-reality, ideas and imaginations are "brought to life" in a digital reconstruction.

The *Harvard Gazette* article entitled "Hawking at Harvard" provides this information regarding black holes,

> "To understand whether that information is in fact lost, or whether it can be recovered [in/from a black hole], Hawking and colleagues, including Andrew Strominger, the Gwill E. York Professor of Physics at Harvard, are currently working to understand "supertranslations" to explain the mechanism by which information is returned from a black hole and encoded on the hole's "event horizon."

The "information" as discussed earlier in this book, is non-physical in origin, but also informs the shape/effects of physical reality. The methods that astronomers use to come up with properties of the black hole are through collecting data (information) gathered from the far reaches of space, and reconstructing the data (information) here on earth. This form of reconstruction lies at the heart of the bridge between the supernatural and natural realms. Enter the discussion of CERN once again.

In around 2009, when CERN and the LHC hit the public newswire, there were expressed concerns with its potential consequences of conducting experiments of such magnitude. Among the concerns was the fear that CERN would spawn *black holes*. From the CERN website:

> **Microscopic Black Holes**
>
> Another way of revealing extra dimensions would be through the production of "microscopic black holes". What exactly we would detect would depend on the number of extra dimensions, the mass of the black hole, the size of the dimensions and the energy at which the black hole occurs. If micro black holes do appear in the collisions created by the LHC, they would disintegrate rapidly, in around 10^{-27} seconds. They would decay into Standard Model or supersymmetric particles, creating events containing an exceptional number of tracks in our detectors, which we would easily spot. Finding more on any of these subjects would open the door to yet unknown possibilities.[26]

Without revealing here whether they had actually achieved microscopic black holes, they professed great knowledge of how it all works. To connect one dot to another, consider the fact that CERN analyses their data through something they call "digital reconstruction." It was recently reported that CERN has made 300 Terabytes of data available for the public to wade through. On their own website, CERN says that they are "digitally reconstructing" the data retrieved over the course of the last few years. Given the probable fact that the LHC does create brief black holes, and with the recent interpretation that black holes could be portals into alternate realities, is it possible that the digital reconstruction will yield a virtual reality that we did not actually create? Such a possibility would confirm alternate realities, accessible through the virtual platform. And is it possible that the reason why Jesus will have to bring about a New Heaven and New Earth is because these openings will cause great chaos for creation? Is the creation being corrupted in its fabric the way humanity was

---

[26] Retrieved from http://home.cern/about/physics/extra-dimensions-gravitons-and-tiny-black-holes.

corrupted by the Nephilim "coming into the daughter of men?" We know from what the scripture teaches that the creation already is "damaged goods". But the damage done may still be getting worse. CERN may be generating more chaos as it operates day-by-day.

CONCLUSION

In conclusion, we have seen that there are curious connections between CERN, the ancient Tower of Babel, black holes, and virtual realities. These entities while different in kind, location, and essence all give rise to massive questions about what humanity is building, why it is doing so, and who might have architected the structures and overseen their construction during the course of developing such technologies and structures. The implications of ancient texts suggest that the "Watchers" or other non-physical, supernatural/spiritual intelligences guided the process. And while this idea reverberates through ancient mythologies, it is remarkable to consider the possibility that something quite similar (if not even crazier) is happening in our day! Science fiction no longer titillates as much as scientific fact. Reality is undergoing revision still and thus we shouldn't be surprised when more changes, even greater in scope, are announced to the world that scarcely understands or appreciates what havoc has already been wreaked on the Cosmos.

We may never know this side of the Second Coming what was going on "back in the day" (circa Genesis 11), or how things will unfold in these highly strange times in which we are living now. But we can be certain that God remains in control and that He will overrule whenever necessary the enemies' stratagems to achieve God's perfect will.

No doubt you will agree with me that the prophecies of Jesus regarding the end times just ahead ring oh so true:

> *And except those days should be shortened, there should no flesh be saved: but for the elect's sake those days shall be shortened.* (Matthew 24:22)

Whether in ancient times on the plain of Shinar or today's incarnation on the border of Switzerland, the *Towers of the Technium* always offer a vacuous promise of a new era for humankind – a way to "make a name for ourselves" – a means to transform humanity into the demigod caste, entities that exceed what God made men and women to be.

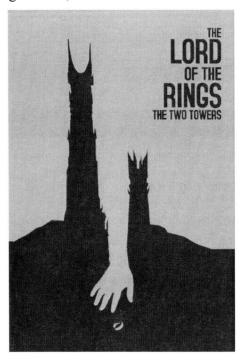

Figure 53 - **TOLKIEN'S *TWO TOWERS***

However, in the end these two towers as in the Tolkien's *Lord of the Rings* (*Orthanc* and *Minas Morgul* being the two towers so labeled in his novels) could be renamed "*Chaos* and *Calamity*" for, all things considered, that fate constitutes all that they portend despite the massive effort humanity expended in creating each of them.

God is not mocked. He knows their schemes and laughs at their boasts. He will crush the new "tower" in Geneva just as decisively as He dealt with its ancient archetype in Mesopotamia, where He reigned down fire upon it, and demolished it leaving a remnant hidden beneath a mountainous heap of bricks. Nimrod did NOT deal God a death-blow 4,000 years ago. He will not harm Yahweh, His Son Y'shua, or His mighty ones – His saints – when the battle is joined in the days not too distant from now.

In the final analysis, the plea submitted by the makers of the Towers of the Technium will ultimately be uttered faintly if altogether inaudibly in the grandest of shame: *Nolo contendere*. "No contest".

ADVANCED CIVILIZATIONS AND THE REAL STAR WARS?[1]

S. Douglas Woodward

WHEN SCIENTISTS COULD BELIEVE

ONCE UPON A TIME, EVEN THE LATE GREAT COSMOLOGIST CARL SAGAN COULD OPENLY CONJECTURE ABOUT INTELLIGENT LIFE WITHOUT WORRY IT COULD GET HIM FIRED. THIS WAS 1966, when he was a professor of astronomy at Cornell. At that point, he wrote a book with a provocative title, *Intelligent Life in the Universe* with a Soviet astronomer, I.S. Shklovskii, a member of the Soviet Academy of Science. It was a challenging time to team with anyone from behind the Iron Curtain since it was at the height of the Cold War. The fact they jointly speculated Mars' two moons—*Phobos* and *Deimos* were not naturally formed moons—no doubt also raised a few eyebrows.

Shklovskii and Sagan asked, "Could Phobos be indeed rigid, on the outside—but hollow on the inside? A natural satellite cannot be a hollow object. Therefore, we

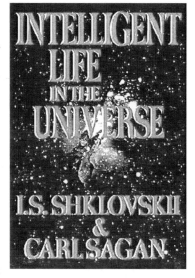

Figure 54 – SAGAN AND SHKLOVSKII, 1966

---

[1] This chapter has been revised and updated for this book from its original publication in *Lying Wonders of the Red Planet*.

are led to the possibility that *Phobos*—and possibly *Deimos* as well—may be artificial satellites of Mars."[2]

Manufacturing moons is no easy business, even for superpowers. To contemplate such things had some scary implications. Whoever made those moons must have vastly superior capabilities and thus, there could be a frightening threat facing Planet Earth. The fact that a U.S. and Soviet scientist came together to write such a book when it was politically incorrect even to be seen in the same venue, much less share a geopolitical opinion, raises intriguing questions: "What could cause the U.S. and Soviets to find it imperative to race to the moon when they were about to blow each other up? It had to be something big. Were they worried that ET was a threat to the human race? Should the Soviets and US ally to protect our planet?"[3]

By 1980 when Carl Sagan wrote *Cosmos*, he had toned down the rhetoric about life existing elsewhere. Despite writing the fictional book (later made into a 1997 movie, *Contact*), he was much more publicly skeptical, not just about life on Mars, but life most anywhere else.[4] It seemed scientists had forever learned their lesson. The fall from grace of Percival Lowell (whose beliefs about Mars were later seen as foolhardy) was not to be forgotten: never talk about life on other planets… people develop unreasonable expectations. Likewise, the example of

---

[2] Carl Sagan and I.S. Shklovskii, *Intelligent Life in the Universe*, Holden-Day, 1966, p. 373.

[3] One of the reasons that Kennedy was assassinated, according to some Kennedy conspiracy aficionados, is that Khrushchev and Kennedy had agreed they would work together to reach the moon. Khrushchev's son, as I documented in *Power Quest, Book Two*, confirmed they reached this agreement. This alliance represented a threat to American contractors in the Military-Industrial Complex.

[4] As authors William Sheehan and Stephen James O'Meara said regarding the "tests for life" executed by Viking I, "For Sagan, the results were depressing. A few scoops of Martian soil, and *poof!* Mars was once the advent of the Space Age (in fact, just up until the Viking Mission), the universe had been a playground for all manner of imagined life. Lowell had his intelligent race of civilized Martians, while Flammarion believed that even Venus and the other planets had inhabitants. In the 1700s William Herschel believed that extraterrestrials might live inside the cool core of the Sun, looking out at the universe through sunspots… Ironically as technology brought us ever closer to achieving the summit of our "mountain Mars," our visions of life on the planet diminished proportionally. It was as if we were looking for life through the wrong end of the telescope." (Sheehan and O'Meara, *The Lure of the Red Planet*, Amherst, New York: 2001, p. 287).

the French astronomer and spiritualist Camille Flammarion, taught the public it quite unwise to mix *planetary science with the paranormal*. Speaking about ANY life beyond our globe was – well – crazy. Look what happened when average folk listened to Orson Welles' 1938 fictional radio broadcast that the Martians had landed. Welles easily lit the fuse of panic. All things considered, scientists should keep their mouths shut about what REALLY might be out there. There truly is no upside.

Subsequently, the experts grew so cautious when talking about the possibility of life on other worlds that the experiments to test for life on Mars, although first thought a success, were later rejected as inconclusive. In a book published in 2003, *The Microbes of Mars*, Authors, Barry E. DiGregorio, Dr. Patricia Ann Straat, and Dr. Gilbert V. Levin discuss the scientific tests for life aboard the Viking I spacecraft, our first soft lander on the Martian surface (1976). Levin was the principal designer of several tests. Despite showing positive results for the existence of life, due to a growing consensus (later reversed) that Mars had no water, test results performed on the Martian surface were dismissed. Not to be forever thwarted, Gil Levin released a study in 1986 and then another statement in 1997 to reopen the issue. He asserted, despite arguments to the contrary, that to his satisfaction the tests carried out by Viking 1 proved living microorganisms existed on the surface of Mars.

DiGregorio quotes Levin as follows:

> The failure to pursue NASA's highest priority (the search for life in the solar system), and the goal NASA once described as "probably the greatest experiment in the history of science," cannot be logically explained. It results from NASA's fear of finding out that its original conclusion about Viking was wrong, supplemented by philosophical and religious elements who insist, for non-scientific reasons, there can be no life elsewhere but Earth.[5]

And I would add, the politically incorrect pressure to avoid making the same mistakes that Lowell and Flammarion made. Conjecturing about life on Mars at NASA could be a career-limiting move.

---

[5] Barry E. DiGregorio, *The Microbes of Mars: A 2011 Addendum to Mars: The Living Planet,* Middleport, New York, Barry E. DiGregorio, Kindle Version, Location 217.

But the point here is not to join the debate—rather, it is assert that even if life was discovered on Mars, we would likely not be allowed to know that fact. Scientists believe for a variety of reasons there is no advantage in getting the populace stirred up about life elsewhere. Likewise, public officials regard the public to be so panicky that the cat better NOT come out of the bag. It is best kept secret.

Dr. John Brandenburg in his book, *Life and Death on Mars*, offered the same assessment—that NASA is prejudiced against ever finding evidence of life on Mars. "This reflex rejection of 'life-as-never-the-simplest-hypothesis' became ingrained in Mars science. Just about any hypothesis, no matter how arcane, will be entertained at a Mars conference, as long as it does not involve biology or the conditions conducive to it. This mindset has continued since the Viking Life experiments in 1976 and led to a crisis in science."[6] The ghost of the Vatican verdict against Galileo still seems to haunt science.

The biggest and best kept secrets, however, involve much more than mere microbes on Mars. Scientists won't talk much about them. Yet, writer/researchers (whose career is not directly tied to funding from grants and professorial peer pressure) are willing to investigate these best-kept secrets. This group (which includes yours truly) are generally denigrated as "alternate historians" and "conspiracy theorists." Despite developing a persecution complex, this group labors on. This irrepressible contingent finds bushels of facts—and even an occasional scientific paper—to back up the story. These alternative theories are not always right, but they still deserve serious consideration instead of being dismissed out of hand.

THE SECRET SPACE PROGRAM

One of the more intriguing topics that these alternative historians and conspiracy theorists address concerns whether the race to the moon between the Soviets and the Americans was "for real." Debates have raged for years (and still do in some quarters) that we didn't really put

---

[6] John Brandenburg, Ph.D., *Life and Death on Mars: The New Mars Synthesis*, Kempton: Adventures Unlimited (2011)

men on the moon. For one thing, the dubious assert we lacked the technology to do it. Then there is another factor: these same "doubters" assert that human beings could not pass through the Van Allen Belts (now known to be three radiation blankets—shields really—that help protect Earth from cosmic radiation). Be that as it may, the matter goes far beyond the moon landing.[7] The mystery is how we acquired so many advanced technologies in the 1950s and whether aliens from another realm in fact threatened us. Several best-selling authors take up this conjecture under the collective moniker, *The Secret Space Program.*

One of these authors and researchers, Joseph P. Farrell (whom this author has cited on many occasions), discusses this so-called "Secret Space Program" and its implications across several of his tomes. In particular, Farrell's 2013 book, *Covert Wars and the Clash of Civilizations*, has to do with global "oligarchs" and how they financed advanced technology development "off the books." Untold fortunes would be essential to pull this off. Perhaps secret assets in the Far East were leveraged (caves full of gold may be hidden in the mountains of the Philippines—a grand story in its own right). The effort demanded creating technologies sufficiently advanced to neutralize the threat of ET. In other words, enormous effort has been expended for over five decades to demonstrate to ET that if they dared attack Earth, they would pay a heavy price.

Figure 55 –JOSEPH P. FARRELL

On the surface – that is, the overt "space race" – constituted a high stakes competition between the U.S.A. and the U.S.S.R. The covert race, however, required "Earth" (the combined forces of humanity)

---

[7] Please note: this author remains fully satisfied that Neil Armstrong and Buzz Aldrin did walk on the moon in July, 1969.

build up enough real technical capability (or apparent capability for bluffing the enemy) that we could persuade would-be alien adversaries to judge us invulnerable to attack. Thus, if the overt space race was for high stakes, the covert space race was for even higher stakes – our leaders determined it was a matter of life and death for the entire human race. With the following analysis, Farrell recaps the storyline:

> Examining the UFO phenomenon and the USA's space program, many researchers, including this one, have advanced the hypothesis that there must be a secret space program, with hidden and very advanced technologies and agendas.
>
> It is one thing, however to maintain this hypothesis, but it is quite another thing to maintain that the needs and requirements of such a secret space program led to the development of an entire "state within the state," as it did in Nazi Germany's case with the SS, or subsequently (and on an even larger scale) in the United States of America, with the creation of vast intelligence, covert operations, and military black projects bureaucracies to deal with the twin threats of Soviet Communism and the more long-term threat of the UFO.
>
> As I argued in the previous book of this small series, *Covert Wars and Breakaway Civilizations,* all this in turn required the creation of enormous and completely hidden and secret systems of finance.[8]

Secret finance and the creation of a hidden branch of government, analogous to Himmler's SS—all to fight a war that the public did not know was taking place. Conspiracies get no better than this!

THE TRIGGERING EVENT

Although Farrell is not unaware that most experts doubt the testimony of Lt. Col. Philip Corso (and the claims from his controversial book, *The Day after Roswell*, 1997 – a book this author feverishly read when first published), Farrell nevertheless sees more than a kernel of truth in the colonel's story (pun reluctantly admitted). Corso explains why he tells his tale with these stimulating words:

---

[8] Joseph P. Farrell *Covert War and the Clash of Civilizations, UFOs, Oligarchs and Space Secrecy*, Kempton, Adventures Unlimited, Kindle Version, 2013, Location 108.

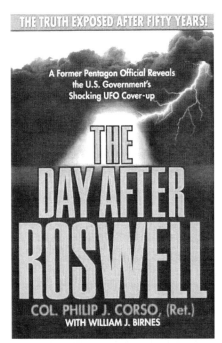

Figure 56 – THE DAY AFTER ROSWELL

The full story behind the SDI [Strategic Defense Initiative, aka "Star Wars" program of Ronald Reagan] and the way it changed the Cold War and forced the extraterrestrials to change their strategies for this planet is a story that's never been told. But as spectacular and fantastic as it may sound, the story behind the limited development of the SDI is the story of how humanity won its first victory against a more powerful and technologically superior enemy who discovered, to whatever version of shock it experiences, that there was real trouble down on its farm.[9]

What Corso said was spectacular enough—a covert war is being fought between Earth and the "aliens"—a war supposedly managed by the U.S. military. Yet, what he said in passing was also stunning in betraying the cosmology behind the war. The "earthlings" were now fighting back against the threats posed by aliens. In Orwellian motif, the farm animals were rebelling against the "owner" of the animal farm.

So were key members of the U.S. government operating from the perspective that humankind was birthed or at least nurtured by aliens? Were the slaves revolting against their masters?

For the sake of argument, we assume Corso supplied an accurate account of how the U.S. government captured "foreign-made" technology when the world's most famous flying saucer stopped flying and crashed in 1947 outside of Roswell, New Mexico.[10] Corso was named the primary

---

[9] Col. Philip J. Corso, *The Day After Roswell*, New York: Simon & Schuster, 1997, pp. 249-250.

[10] In using these words, I'm accounting for the fact, as Farrell himself argues, that the technology in the 1947 Roswell crash may have been manufactured by Nazis (rather

military officer-in-charge of taking this "Roswell technology" and exploiting it for the benefit of the United States. That is, Corso became responsible for privatizing these advanced technologies captured from the Roswell crash site – wherever they originated – and putting them into play for practical purposes. His job was to place them in useful applications, whether for military designs or private use.

It wasn't until years later, however, that we might have discovered what was going on, when President Reagan opaquely spilled the beans about the U.S. and the Soviets working together – sharing the technologies created by the Strategic Defense Initiative (aka SDI, affectionately labeled "Star Wars"), one grand day in 1987 when speaking at the U.N. In his speech famous to UFOlogists, Reagan hypothetically conjectured the Cold War would end the instant we discovered that we shared a mutual enemy from outer space! Should we figure out that ET was eyeing our world (in the manner told us 90 years earlier when H.G. Wells wrote his novel, *War of the Worlds*), there would be little doubt it would put into perspective how much we held in common with our mortal enemies, the Soviets.[11] Of course, Reagan might have only been talking hypothetically. But the genie was out of the bottle even if that genie really was just a vapor.

Recall SDI's ostensible purpose was anti-missile defense. But the ultimate reason, if we caught the implications of Reagan's chat, went far beyond creating a missile defense shield to nullify the use of "strategic" nuclear weapons (i.e., inter-continental ballistic missiles, ICBMs). With SDI deployed, not only would a surprise attack be rendered effectively impossible, it would eventually spell the death of such weapons altogether. Moreover, since the "accelerated particle beam weapon" of SDI could shoot down approaching enemy missiles, it could also (more importantly) shoot down flying saucers! Finally,

---

than aliens as Corso asserts), who escaped Europe after World War II and set up shop somewhere in the Americas.

[11] General Douglass MacArthur was much more emphatic, as he was prone to be, at a WestPoint speech in 1955: "The next war will be an interplanetary war. The nations of the earth must someday make a common front against attack by people from other planets." In 1955, the U.S. Military was already thinking alien attack!

according to Corso, by the time he wrote his book in 1997, SDI had already done just that. A UFO or two had been shot down. At least that's what we've been told.

Perhaps it was no surprise that only a few years after Reagan's speech (with SDI now proven against the "real enemy"), the Cold War ended. Was it coincidental that at the same moment we effectively neutralized ET's threat (to the satisfaction of certain key government officials within the world's only two superpowers), the Cold War thawed out?

A VAST INSTITUTION – INVISIBLE & UNGOVERNABLE

For decades before Corso wrote his book, according to Farrell and others who subscribe to the notion of a Secret Space Program, there were myriad clandestine duties carried out by, note carefully, non-U.S. military personnel. Efforts involved far more than the singular Roswell-connected actions managed by Corso. In fact, the lion's share of projects dealing with ET's threat was a black-ops program run by scientists and militarists outside the control of our government (or any official governing body foreign or domestic).

In his books, Farrell goes into considerable detail explaining that the strategy involved hiding an anti-alien defense initiative within the conduct of the Cold War. He borrows a seminal phrase coined by author and UFO researcher, Richard M. Dolan from his multi-volume work, *UFOs and the National Security State* (Volume 1 published in 1973) who argued that a "breakaway civilization" had taken charge of the overall program. This civilization was *breakaway* because any known public governance did not control it. It may have begun under the auspices of major governments. But once it overcame bureaucratic inertia, it became impossible to reign in. The breakaway civilization began with many of the best and brightest working together across national boundaries, perhaps for the protection of "Mother Earth." But according to these researchers. it morphed into a Frankenstein monster fully untethered, primarily protecting the interests of globalists and "international elites." It lurks in the dark with no visible headquarters, and beyond contacting.

According to Dolan, Farrell, and others who contend this civilization exists, there are identifiable persons who "interface" with this covert set of out-of-control institutions. While Farrell suggests who might be involved—that is of no particular concern to us here. The issue that is pertinent: *a covert war exists, it involves advanced technologies, it is being fought against extraterrestrials*, and we are "learning a lot more as we go forward." Not only are we fighting the future (the final outcome of the administered plan by the power players—as the first "X-Files" movie was sub-titled), as we progress along the way, we marvel at the additional insights uncovered concerning the history of ancient civilization on our planet, our solar system, and even the galaxy we call home.

Perhaps we should pause here to catch our breath for a moment and to recap what we have covered. We have seen several examples of key persons who stated aliens were at the door, the evidence indicates they are powerful enough to build moons, and then, these alarmists went silent. The most vital effort in the history of mankind went off the radar and a "Secret Space Program" was unofficially (but factually) put into action. How was this institution financed? That too is a grand story told by Farrell, but we haven't the time or space to trace his research here. Suffice it to say, financing was accomplished by hidden gold buried in the Philippines, untraceable bearer bonds carried about by clandestine agents, and remains to this day an "off the books" enterprise known to only those initiated by Farrell, Dolan, and those initiated by their writings.

THE COSMIC WAR—A REAL STAR WAR?

Indeed, I have outlined so far, however, only the first portion of the colossal "mythos" of Joseph P. Farrell. It now seems justified to consider Farrell's take on *Ancient Alien Theory* because, simply put, his views go in a very different direction from "standard theory" and in many ways ring far truer. His perspective also comprises a more frightening narrative. Moreover, it entails a much more complex tale than the standard theory. And as we are about to see, it comprises a vast, multi-layered saga. We should also note: some aspects of Farrell's take on "Ancient

Alien Theory" resonates with the Bible's account – at least as explored by the late David Flynn – regarding cosmic powers and their ulterior motives. [12]

Therefore, we turn next to what Joseph P. Farrell believes comprises our most essential history with ET.

As reprise, *traditional* Ancient Alien Theory (admittedly, a strange word to use in this context) holds that ETs are our space brothers. They have come to Earth to keep us from blowing each other up given our misappropriation of nuclear energy. It seems WE are a race that declares war at the drop of a hat. THEY are far more civilized. THEY settle their disputes, so we are told, without warfare. The story of Mars follows the same plotline. From the writings of the nineteenth century to the "Marxist" utopian theories of the early twentieth century, the intelligentsia has argued that Martians were older and wiser than we, their younger and less socially advanced humanoid siblings. We could learn from THEM. Indeed, we must learn from THEM if we are to survive and evolve to the next level.

In current Ancient Alien Theory, humankind's religions are regarded as early attempts by simple-minded people of ancient times to come to grips with alien visitors. Our connection to ET is, however, most intimate. Aliens supposedly tampered with our genetics, improving our genome. According to many Ancient Alien theorists, especially Zecharia Sitchin and his followers, we are today already chimeric; that is, we are *hybrids of alien and Homo sapiens ancestry now*. We are way past mere mentorship—we are into motherhood amidst modernization.

In our day, however, ET still continues to be maintaining his technological advantages—he remains much more advanced than we. Those that believe in the Gospel of ET maintain that in every sense, aliens are grander, more peaceful, and have mature values. Now ET stands ready to return, to facilitate our transition to the next phase of evolution.

---

[12] For a detailed study on Mars and David Flynn's theories, please see Woodward's book *Lying Wonders of the Red Planet*, Oklahoma City: Faith Happens, 2014.

Of course, advocates ignore any evidence to the contrary—and there is *considerable contradictory evidence*. Proponents overlook alien abduction victims who testify of brutal experiments, harvesting human eggs and semen, gestating hybrids, kidnappings, and unwelcomed nighttime visitations. All of these behaviors are dismissed or downplayed by advocates for the "Space Brother" point of view. No old hags, no black helicopters, no aliens at the foot of our beds – only good things.

Figure 57 – THE COSMIC WAR

With Joseph P. Farrell's perspective, however, the benign alien postulate gets tossed out the window. The answer to the challenge of "from whence intelligent life hails," appears derived as much from beings who have been resident on earth for millennia (perhaps ten to twenty millennia) as it does from locations beyond our earthly realm. The point is not so much that ET is not coming someday soon—he is here and influencing our civilization for millennia—if not millions of years.

For Farrell, the elite on this planet may be descendants of intelligent beings originally *spawned* (I use that word intentionally) NOT "a long time ago in a galaxy far, far away," but more likely from a planet within our solar system—namely, the missing planet postulated by Dr. Tom Van Flandern, the late head of the Naval Observatory, who reawakened the notion that there is a missing planet right smack dab where the asteroids are. This is no coincidence. This elite has been operating literally for aeons, from a point in the past Farrell depicts as "high antiquity" – a time long, long ago for which he also coins the term "paleoancient." The mythos builds on a nation that creatures near to us (hear on this earth or perhaps on Mars), maybe even our DNA kin, our forebears, had built a vast and highly technical civilization. In other words, like classic Ancient Alien Theory made popular by the History

Channel during this past decade, Farrell's proposal combines a lost advanced civilization with activities occurring in outer space – but on a scale that far exceeds the traditional theory.

From this author's perspective, that does not necessarily mean that Farrell's mythos is more likely to be true than the traditional ancient alien theory – but it stands apart as a far more interesting alternative since it accounts for anomalies left out of the standard theory and may be better reconciled (not perfectly but substantially) to the biblical record. But what is the evidence for this connection?

First, Farrell argues that various physical realities of our solar system are better explained, not by natural processes, not just by cosmic catastrophes, but by a cosmic war – a true "star wars" that encompass the entire galaxy,[13] enlists the use of galactic energy, and places the Earth at the center of the conflict.

With Farrell, we encounter the exploding planet hypothesis of Tom Van Flandern once again, but in his scenario the annihilation does not result from a random collision with another world—as the reader may recall in the theory of Immanuel Velikovsky (1895-1979, made famous through his 1950s' book, *Worlds in Collision*). Instead, the explosion results from a powerful "planet busting beam" (think the "Death Star" in *Star Wars*). Here, as per usual in his books, Farrell explains the science behind his theories – which I won't delve into here with much detail. However, at the risk of losing some readers, we must digest a few technical elements not only to satisfy the curiosity of those readers up to the task, but at least to appreciate the broader strokes of Farrell's theory.

Farrell begins by ruling out various explanations that do not have the inherent energy to bust a planet. He explains no known example exists of natural processes exploding a planet. An asteroid impact would not generate enough kinetic energy to do the trick. Nor would a mere incident involving the (natural but rare) occurrence of *nuclear fission*

---

[13] I will not discuss an element theorized by Farrell and a few others, that the "pulsars" (those supremely fast rotating stars detected by radio telescopes) are non-randomly placed in the galaxy. They could serve as a network of beacons for interstellar travel. In effect, beings that do interstellar travel, assuming for a moment that such beings exist, would navigate by referencing this set of pulsars.

be sufficient to blow up a planet. To destroy a planet, we need a powerful explosive, one that transcends the limits of our know-how. Exactly what are we talking about? So get ready to have your mind blown.

It is most significant that high levels of Xenon 129 exist on the Martian surface – three times the levels known to be present on any other planetary body. So while Farrell makes the point conditions could randomly coalesce to allow for spontaneous nuclear combustion, such conditions would not produce the levels of Xenon 129 we detect on Mars. Consequently, according to Farrell, we know of no model that allows for a massive nuclear reaction capable of exploding a planet. And yet, Mars appears to have been affected by some form of nuclear blast. How could this be?

Farrell postulates that proximity to an exploding planet, done in by an engineered explosion (making use of a form of nuclear fusion we will discuss next) could very well explain the presence of this isotope on Mars. On this point Farrell cites Van Flandern once again, as the late scientist proposed that active intervention by an "intelligent" agent should not be ruled out as explanation for an explosion big enough to destroy a whole planet. Therefore, Farrell concludes that only a weapon can blow up a sizeable heavenly body. Once destroyed with the weapon he describes (employing this most sophisticated nuclear technology), radiation residue would linger in the vicinity for quite some time. In fact, on Mars we may just have a case of interplanetary nuclear fallout. Mars may be the biggest smoking gun that we ever will see!

ION CANONS IN THE COSMIC WAR

So what weapon could achieve this "mother" of all cosmic cataclysms? In a phrase: a *plasma pinch*. What, pray tell, is a plasma pinch?

To begin with, we must acquaint ourselves with a scientific fact not taught to my generation in elementary school: there are four, not three, states of matter: solid, liquid, gas, and *plasma*. Plasma is a state of matter that only exists inside nuclear reactions. The dictionary in Microsoft Word 2013 defines a plasma as "a fourth state of matter distinct from solid or liquid or gas and present in stars and fusion reactors; a gas becomes a plasma when it is heated until the atoms lose all their electrons, leaving a highly electrified collection of nuclei and free electrons." Forget for now that you might want to buy a plasma television.[14] We are focused here on much bigger (and hotter) plasmas.

The plasma "pinch" more or less conveys the notion of compressing a nuclear reaction into an elaborately structured nuclear containment unit of sorts, squeezing the nuclear activity in such a way that a "longitudinal pulse wave" is created. A *longitudinal pulse wave* is different from a run-of-the-mill latitudinal sound- (or light-) wave. As Farrell explains, a sound wave may be compared to throwing a jump rope up and down. On the other hand, a *pulse wave* may be likened to a *pushing* a yard-stick. Working in tandem with your hypothetical jump rope partner, you expend energy lifting the jump rope up and pulling it back down. Generating a wave in that manner wastes most of the energy just by creating the wave – the wave itself is not that stout in terms of conveying energy to the other side of the wave, to its terminating point. If I'm your partner, I feel the wave but it doesn't move my hand up or down very much. In contrast, a pulse wave would be like pushing the yardstick directly at the person on the other end of the stick. All of your energy, the force you create, is felt directly by your partner. Because you and your partner are connected (in our analogy, you have "addressed" your partner by the connection between you – each holding one end of the stick), therefore, your energy (all your effort) is put to work. If you aren't careful, you might just knock your friend off his or her feet.[15]

In our elegant hypothetical particle beam accelerator (in Farrell-speak, aka *a scalar weapon*), the device allows particular particles to escape at either end. Electrons go out at one end and fused nuclei, aka ions, go out the other (in this discussion, ions are fused protons and neutrons; therefore, ions are separated from their typical electron partners). Importantly, *ions contain the mass of any atomic particle*—as electrons

---

[14] This type of flat panel TV is called a "plasma" display since the technology utilizes small chambers containing electrically charged ionized gases.

[15] Tesla purportedly created a *pulse wave radio* to talk to the Martians!

have no mass. Therefore, being hit by ions (the so-called ion canon we hear tossed about in science fiction television shows and films) packs quite a punch. Targets hit by a sufficiently robust ion beam would be obliterated.

Imagine for a moment harnessing the nuclear reaction of a star and squeezing that energy through our ion canon. Farrell explains how that can be accomplished theoretically (which I won't explain here – I've risked losing enough readers already). Suffice it to say, a star turned into an ion canon would be the ultimate "Death Star".

So how does one aim this radiation beam? Farrell explains how targeting is based on knowing the atomic "address" where your target resides. Aiming the beam requires "connecting" your canon with your target. Remember the yardstick analogy. Once you connect, the force or energy expended on your end can be applied directly to the target on the other. Energy loss may be virtually zero.

Addressing particles or collections of them (even a massive collection as in a planet) can be accomplished by detecting its *resonant signature.*

To explain that concept: understand that atom in the universe has its own unique atomic signature—which amounts to its very own resonance. To impact it, to target the beam at it, one needs to know its signature. I would use the phrase "being on the same wave length" as your target, but that might actually be confusing latitudinal waves with longitudinal waves. It would be more accurate to say being "connected" with your target—your weapon holds one end of the yardstick and your target holds the other. Just like a radio transmitter and radio receiver need to be tuned to the same frequency to work, *the particle beam weapon works when their frequencies synchronize.* To repeat, despite being infinitely larger than a single atom, planets and stars each have their own signatures (aka resonances). Dial in their exact resonance… release the massive particles through your beam weapon… and WHAM! You have destroyed a planet (or a sun if you are really "shooting for the stars" – which brings a literal meaning to that familiar metaphorical phrase).

Of course, it's not really that simple. I skipped quite a few steps here, which reading Farrell for yourself can remedy. However, for our purposes in this chapter, my explanation should suffice. But be warned: according to another expert Farrell cites, Lt. Col. Tom Bearden, if you aren't careful, your particle beam weapon can backfire. This could result in the greatest case of shooting yourself in the foot ever seen in the history of the universe. Bearden cautions:

> If the discharge happens to tickle the Sun and Moon's feedback loops the wrong way, you'll get convulsions of the earth, mighty burps of the sun raining fire and brimstone on the earth, and a violent increase in the

Figure 58 –EXAMPLES OF RONGORONGO PETROGLYPHS

> interior heat of the earth's molten core, with a concomitant eruption of that core right up through the mantle... (Whenever) one activates a large scalar (electromagnetic) weapon, one immediately places the entire earth in deadly peril. The slightest misstep, and it's curtains for everyone. And it's curtains for the earth as well.[16]

---

[16] Farrell, Ibid., p. 47. Citing Tom Bearden, *Fer De Lance*, p. 408.

WHEN DID THIS COSMIC WAR HAPPEN?

If you were stationed on Earth when a cosmic war transpires, and one of the planets orbiting our sun is taken out by an ion canon, you would see enormous aurora displays in the heavens. Plus, these displays would hang around for quite some time. Eventually, everyone around the globe would see the aurora as it produces a persistent magnetically induced image—with one particular shape resembling a humanoid. Consequently, hiding the fact you annihilated a nearby planet would be most difficult—using this scalar weapon would create witnesses everywhere. And if the warfare happened within, say, the last 12,000 years as Farrell speculates it may have, our forebears on every continent[17] would have taken note of the event and recorded it in the artistic medium most readily available to them: image painting on the walls of their cave dwellings. Based on the work of a plasma physicist Anthony Peratt, Farrell conjectures this is exactly what happened. These underground notations were the focus a scientific paper developed by Anthony Peratt while working for the government at Los Alamos Labs.[18] The fact we had a plasma physicist working for the government and doing papers on plasma events dating back to pre-history, raises interesting questions in its own right. We will not have the time to investigate that matter.

Suffice it to say that *our government might know about "paleoancient weapons"* that an advanced ancient civilization used—and *may be seeking to discover the technology behind such planet busting potential—before someone else does.* Farrell pays particular attention to ancient texts and the so-called Tablet of Destinies. This record (a tablet made of some type of resilient substance) supposedly contains the secrets

---

[17] Were cavemen Homo Sapiens? Were they annihilated in this Cosmic War? Were Adam and Eve successors, that is, Homo sapiens sapiens modeled after their predecessors but given a genetic upgrade by Yahweh that advanced intelligence and the capacity to commune with Yahweh? All good questions. No firm answers.

[18] Anthony Peratt's paper proposes a connection between plasma displays in the atmosphere with ancient petroglyphs. His paper may be found at the following link: http://www.scribd.com/doc/14145750/Anthony-Peratt-Characteristics-for-the-Occurrence-of-a-HighCurrent-ZPinch-Aurora-as-Recorded-in-Antiquity.

of this "scalar" technology.[19] Farrell suggests it may still be buried in the sands of the Middle East and might have been the true "weapon of mass destruction" that America sought when it invaded Iraq. If not in the ancient land of Sumer/Babylon, Farrell suggests this ultimate recipe book for destruction could be buried on the moon or Mars! Perhaps. Then again Farrell may just have a flair for the dramatic.

To flesh out the speculation a bit more: Human beings saw this stunning display in the skies – an aurora borealis surpassing any semblance of what they had seen before. Additionally, they may have been physically impacted by this plasma event (by feeling great heat or experiencing resultant earthquakes). Regardless of the physical threat (which ultimately may have destroyed ancient civilization on earth), the troubled population recorded these awesome sights worldwide in petroglyphs where they dwelled. To be clear, caves might not have been their normal abode. They may have taken to the caves for protection (shades of Revelation 6 and the opening of the sixth seal!)

Figure 59 – EASTER ISLANDS MOAI

Another scientist, Robert M. Schoch, corroborates the worldwide witness of this major plasma phenomenon. The reader may recall it was Schoch who famously dated the Egyptian Sphinx to a much earlier date in antiquity than conventional history allowed (dating this signature icon of Egypt all the way back to 7000 BC). Schoch's archeological investigation demonstrated the Sphinx showed signs of rain erosion – which could only have occurred during a much wetter time –

---

[19] Farrell guides the reader for a lengthy analysis (too long for my taste) of the Edfu texts, Egyptian *Book of the Dead*, and the Sumerian *Enumu Elish*,

9000 years ago when it could rain torrents in North Africa. Besides his studies at the Giza plateau, Schoch has researched ancient civilizations in various corners of the world, including the indigenous people of Easter Island in the South Pacific. Schoch studied the famous *Moai* and images of the *Rongorongo* petroglyphs. According to Schoch, images exist in that pictorial language that match those cited by Anthony Peratt. Schoch discusses this connection and the great plasma event of 9700 BC on his website:

> Oddly, the indigenous Easter Island Rongorongo script may hold the answer. But first we have to consider the concept of the fourth state of matter – plasma. Plasma consists of electrically charged particles. Familiar plasma phenomena on Earth today include lightning and auroras, the northern and southern lights, and upper atmospheric phenomena known as sprites. In the past, much more powerful plasma events sometimes took place, due to solar outbursts and coronal mass ejections (CMEs) from the Sun, or possibly emissions from other celestial objects. Powerful plasma phenomena could cause strong electrical discharges to hit Earth, burning and incinerating materials on our planet's surface. Los Alamos plasma physicist Dr. Anthony L. Peratt and his associates have established that petroglyphs found worldwide record an intense plasma event (or events) in prehistory.
>
> Dr. Peratt determined that powerful plasma phenomena observed in the skies would take on characteristic shapes resembling humanoid figures, humans with bird heads, sets of rings or donut shapes, and writhing snakes or serpents – shapes reflected in countless ancient petroglyphs. The Easter Island rongorongo script, recorded on antique wooden tablets, is composed of similar shapes as the petroglyphs. Studying them in detail... I concluded that the Easter Island rongorongo tablets (the surviving tablets are copies of copies of copies...) record a major plasma event in the skies thousands of years ago. This, I believe, was the event that brought a final close to the last ice age. [20]

Schoch does not connect the great plasma event with Farrell's Cosmic War. But his research at Easter Island at least confirms a cataclysmic event took place and was registered by an intelligent race of beings who painted their observations and used them to adorn the walls of their homes (i.e., caves) all around our globe during a timeframe that

---

[20] Citation: Robert M. Schoch: http://www.robertschoch.com/plasma.html.

might correspond to a destruction of all earths' life forms explicated by the so-called *Gap Theory* held by many biblical theists.[21] Could this same event have been the obliteration of Mars' planetary partner? Could this have led to the destruction of a possibly-not-so-mythical Atlantis as many speculate? For Farrell, the timing of his Cosmic War ranges from millions of years ago to a time much more recent—such as an event 12,000 years ago, circa 9700 BC. In his book, *The Cosmic War*, Farrell considers the timing:

> Once this highly speculative concept is entertained [of a Cosmic War], it opens the door to a resolution of other chronological issues, for it allows the war to have occurred at any stage that such a society [here on Earth] might have emerged. In short, and barring the consideration of other types of evidence for the moment, the door is open for the cosmic cataclysm to occur anywhere from millions, to mere *thousands*, of years ago [emphasis mine]. And as has already been seen from the evidence presented in this chapter, there are two loci around which a chronological resolution must be orbited: on the one hand, it must account for the existing planetary data of such a catastrophe, from the asteroid belt as remnants of a missing exploded planet in our solar system, to the electrical discharge scarring on the various moons and planets—most notably Mars – in our solar system [The famed scar is the four-mile deep Valles Marineris on Mars' surface, the length of which is about 3,000 miles or the length of the United States from New York to San Francisco]. The evidence necessitates a much later dating [closer to today], for once one adds intelligent observers into the mix to observe and record these events, one perforce cannot be dealing with the primordial conditions of the solar system… the petroglyphic evidence compiled by [Anthony] Peratt, and the textual evidence of the myths themselves, fix another terminus a few thousand or tens of thousands of years ago. [22]

Despite the argument he advances for a planetary explosion represented in humanity's "cave paintings" referenced as the "Problem of Peratt," (which means that the evidence he evinces demands there

---

[21] Referenced earlier, The Gap Theory conjectures that between Genesis 1:1 and 1:2, there was a gap of thousands, millions, or billions of years. The current earth was not an event of creation, but re-creation, reclaiming an earth that had been turned into a chaotic wasteland, perhaps as a result of Lucifer's rebellion.

[22] Joseph P. Farrell, *The Cosmic War: Interplanetary Warfare*, Modern Physics, and Ancient Texts, Kempton: Adventures Unlimited, 2007, pp. 78-79.

were sentient, intelligent beings who observed the event). As an academic hedge of sorts, Farrell does not consistently argue that this event was this proximate to our day. In fact, he may be slavishly reliant upon Van Flandern's proposed timeline for a cosmic event involving Mars, and not just one but two exploding planets. The problem with Van Flandern's timeframe: it does not allow for humans to be advanced to the point where they could witness and record the event. Unless intelligent humans existed on Earth over 3 million YBP (years before present).

Assuming a traditional evolutionary scenario for the sake of Farrell's argument, humanity did not awaken to abstract thinking until the last 100,000 years or so. Van Flandern had suggested the timing of a second exploded planet was 3.2 million years ago (he proposed a first planet may have blown up 65 million years ago related to the now-accepted catastrophe on earth that ended the dinosaurs' reign). Farrell defaults to the 3.2-million-year timeframe through of the rest of his book.[23] This includes a positive review of Michael Cremo's "hidden history of the human race" – an unusual *alternative* point of view contending that a sophisticated humanity, capable of abstract thinking, has been resident on earth for millions of years. Now that perspective comprises a true minority report. Still, it constitutes one way to reconcile cosmic events with human artifacts.[24]

---

[23] Farrell comments that if Van Flandern's "original Planet V supported intelligent life, then Mars may have been inundated with debris of a very different, artificial, nature." Ibid., p. 22. As to Van Flandern's thoughts on whether there was intelligent life on Mars itself, he "believes that these structures, if artificial, were built by some civilization prior to the event at 3.2 million years ago." Farrell here quotes Tom Van Flandern's book, *Dark Matter, Missing Planets, and New Comets,* p. 435.

[24] A quick note in passing: no one can say that Ancient Alien Theorists are obliged to defend evolutionary theory—they are as prone to call out Darwin as Theists are. As with Theists, for Ancient Alien theorists, Earth is not a "closed system." Consequently, most prefer to keep their feet in both circles depending upon what best suits their purpose.

CONCLUSION: THE COSMIC WAR & ITS IMPLICATIONS

Summarizing the points Farrell makes regarding the *Cosmic War,* including a couple we haven't mentioned yet (in a feeble attempt to streamline my compilation of this rather difficult subject).

First, geological support exists for a "cosmic catastrophe" on Mars; specifically, scars on Mars suggest a planet exploded nearby:

> Mars along of all the planets in the solar system has the best geological evidence for the type of Flood described in the Old Testament and in other ancient legends and traditions. Indeed, the severe hemispherical disparity one encounters on Mars [the southern hemisphere has few craters, the northern hemisphere is covered with them] is exactly explained by the hypothesis, for "one hemisphere would have been heavily bombarded, and the other barely touched by the explosion."[25]

Second, there is the exploded planet itself: Farrell conveys that (1) it was large with the same mass as Saturn; (2) it was solid, for the debris consists of carbonaceous asteroids, and (3) it "was very likely a water-bearing planet, since Mars exhibits definite and distinct evidence of sudden, massive flooding across its entire southern hemisphere."[26]

Mars was permanently scarred by the Cosmic War, but its neighbor was destroyed in a plasma discharge that could be "seen" and felt on Earth. Only briefly mentioned before, we must make a few more points about the mechanism for accomplishing this "Death Star" function. It was, Farrell asserts, recorded on a Tablet of Destinies, and discussed in the ancient texts of Sumer and Egypt that Farrell reviews in *The Cosmic War.* The ultimate trophy in the space race today (now joined by India and China) may be finding that "tablet" and its planet-destroying technologies expounded thereon. In regard to this tablet, Farrell compares what happened in ancient times with the masters who held the tablet, with the Allies at the end of World War II. He summarizes this analogy as follows:

---

[25] Ibid., p. 21. Farrell here cites Tom Van Flandern's book, op. cit., p. 427.
[26] Ibid., p. 14. I believe Farrell has the Martian hemispheres reversed.

The Tablets were inventoried, and some carted off and used elsewhere by the victors in a kind of "paleoancient Operation Paperclip," some were deliberately hidden because of their potential destructive power, and because of the impossibility of destroying them, and some components were permanently destroyed.[27]

There is, of course, no compelling evidence offered by Farrell that proves the Tablets ever existed, other than considering the testimony of the ancient texts themselves. Of course, if a crime was committed, there has to be a motive. And without the hunt for the Tablets and the information they supposedly contained, his theory of the crime lacks a motive – this crucial element goes missing. To play out the analogy: we have a body (Mars and the asteroids); we have a possible weapon (albeit it conjectured – the particle beam weapon aka ion cannon, a planet buster than could leave nuclear residue at the crime site); but the motive is missing. Could it be that the motive is found in the Bible?

Fourth, we must now connect the element not mentioned up to this point in the context of Farrell's mythos. This issue connects the players of the Cosmic War to the issue of the Nephilim – those giant beings that resulted when the *Bene Elohim* came into the "daughters of men" and begat children through them (Genesis 6:4). Farrell supposes that if intelligent beings escaped from the exploded planet to our own, those beings would have been a much larger size, possessing a robust skeletal structure with "muscles to match", in order to deal with the much higher level of gravity they would have experienced there. "In short, such creatures would be, by modern human standards, giants."

No surprise then that Farrell cites the research of noted evangelical author Steven Quayle; specifically, his extensive studies on giants[28] referencing many classic (Greek and Roman) sources. Quayle's work is called upon in Farrell's study *The Cosmic War* to help substantiate Farrell's adjunct theory that these giants arrived on earth sometime in prehistory (readers may recall I discussed this in my book *Lying Wonders*

---

[27] Ibid., p 382.

[28] See Quayle's research at his website: http://www.genesis6giants.com/.

of the Red Planet when discussing the amazing work of David Flynn). According to Farrell's hypothesis, the Nephilim were instrumental in building a "high civilization" incorporating advanced technology and enormous monuments, including such structures of the Pyramids.[29] They were not big, ugly, and dumb. They possessed intelligence far exceeding our own. In fact, the *Book of Enoch* indicates that the Nephilim taught "high technology" (and some low technology too, like cosmetics) to their human apprentices. Yahweh was not pleased the Nephilim provided instruction to his most-prized possession – humankind.

Farrell's theory does not entirely correspond to the views of Quayle, L.A. Marzulli, and the authors of this book, as Farrell interprets the concept of the *Bene Elohim* ("Sons of God") figuratively while we interpret the phrase literally. That is, we believe the Sons of God were fallen angels – perhaps occupying another planet but more likely, originating and residing in a different realm altogether. Farrell suggests instead that these beings were very large humanoids, racially like us (inasmuch that we could beget children together), but who came to Earth as a result of the Cosmic War. Farrell supposes the beginnings of our race may have dated from millions of years ago. Additionally, he even theorizes that our species may have originated somewhere else in the Cosmos.

Nonetheless, while rejecting most of these suppositions, I remain fascinated that Farrell comes as close as he does to the perspective of the authors of this book (which we contend conveys the meaning of the Bible regarding the Nephilim).

Farrell connects another set of dots that, to say the least, adds yet another level of intrigue. Farrell notes the fact Egyptologist and mystic James Hurtak had published his remote viewing experience, "seeing" Mars' Cydonian "face" before the 1976 Viking 1 photo of THE FACE generated so much popular interest. Farrell worries that remote view-

---

[29] When I met Steve Quayle in Oklahoma City back in the fall of 2013 at a dinner with Steve, Gary Stearman, and Bob Ulrich, he was unaware of Dr. Farrell's research and reference to his material. Of course, Steve's essential point of view corresponds to many today.

ers may have been contracted by national governments (or the "breakaway civilization" itself) to search for the secret (perhaps only hypothetical) Tablet of Destinies. He comments:

> Hurtak's viewing of the Mars' Face thus raises another disturbing possibility, one known to be in use presently by the various countries' militaries and also by large corporations researching exotic technologies, and that is that such technologies might be located via such processes. And it raises the possibility that such processes were being used to guide technological exploration of Mars long before probes were actually sent. If so, it casts another shadow on [Farrell's and Dolan's] Two Space Programs Hypothesis. It will only be a matter of time before someone, somewhere, attempts to use the same process to view and generally locate the missing components of *the cause of that Cosmic War, the Tablets of Destinies*. [Emphasis mine, speculation Farrell's]

Should that search be underway and exist as a motive for what continues to transpire in the strange and bizarre world of UFOs and ET, we can only hope that the Tablet remains lost and not found.

Finally, Farrell considers the possibility that a super-intelligent evil personage has been involved throughout the history of *The Cosmic War* and whose threatening presence may lurk somewhere still. [30] If there was such a war, it implies a celestial extent to humanity, or "whomever," and a sophistication of technology and a potential for destruction that we can scarcely imagine. If one adds into that volatile mix the clear indications of our most ancient myths, that there was also an ancient and preternaturally malign intelligence behind the most primordial revolt; that it is a frightening scenario indeed.[31]

This author does believe that such a malevolent intelligence continues to carry out his plan for the deception of humanity, and ultimately, to transform humanity into a form resembling his nature and not that of our

---

[30] Of course, the Bible asserts he does; he has become "the Prince of the Power of the Air" (Ephesians 2:2; in Greek, *"archon exousia aer"*, most likely meaning the "delegated chief authority who exercises his authority in the second heaven, literally our atmosphere, implying his realm of rule lies beneath the third heaven where God reigns supreme").

[31] Ibid., pp. 413-414.

species' actual creator. This story may link to our ancient past, but it surely portends profound connections to our future.

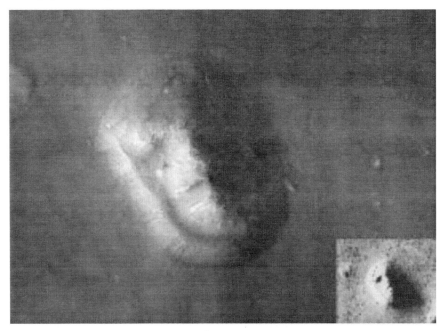

Figure 60 – THE FACE ON MARS

According to *Ancient Alien Theory*, humankind has been altered genetically in ages past by beings other than the God of the Bible.

This generates many more questions:

- Is this alteration likely?
- Is it possible that there lies hidden in us the evidence of past tampering?
- Could it be that this program to alter our genome has commenced yet again?
- Could aliens have actually been present with us all along and we didn't recognize them for who they were?
- Or were these supposed forebears to humanity really just fallen angels the Bible references in Genesis 6 who tampered with human genetics to create beings in their own image as we have expounded earlier?

Furthermore, could the mechanisms for the scarring on Mars not be a particle weapon but something far more "natural", i.e., electric discharges created when objects in the heavens passed by one another so closely that they created an electric arc leaving its telltale sign permanently etched in the Red Planet?

That is the hypothesis of the Electric Universe discussed earlier. It is also a sneak peek at what we intend to discuss in much more detail in the sequel to this book coming in 2017.

CONCLUSION:
THE FUTURE QUANTUM WORLD
Josh Peck

INTRODUCTION

L IKE IT OR NOT, WE ARE HEADED TOWARD A QUANTUM FUTURE. MANY TECHNOLOGIES BASED ON QUANTUM CONCEPTS, ARE UNDER DEVELOPMENT NOW. BEFORE LONG, WE WILL NOT recognize the world we live in. Despite our qualms about the future, as Christians we must adapt to these changes or get left behind. However, when it comes to accepting technological changes – new appliances, new medical devices, new cosmological concepts, even new beliefs about the universe – how can we decide what God finds acceptable and, contrarily, what He would detest? To follow the path allowable to God, we must keep current with what humanity is learning about the universe. As secular knowledge advances, spiritual understanding must advance as well.

Without a doubt, science and technology are rushing headlong into unknown realms. Even though the human powers in charge don't understand the full consequences of their research, scientists often strive unrestrained in their pursuits. On an ultimate quest for discovery, moral or ethical constraints may not interfere much with their methods and might not govern the outcomes.

Perhaps unpredictably, as Christians we might be able to sympathize with the passion that drives scientists – at least to a certain extent. We quest on behalf of Christ's Great Commission (Matthew 28:19-20); and nothing, be it torture nor death, should compromise our assignment. Jesus Christ set forth our mission for the temporal betterment of all humankind and the eternal salvation of each and every soul.

However, therein rests the difference. At best, the goal of scientific innovation apart from God (determined by man alone), consists in the physical improvement or empowerment of humanity; transforming our environment to what the powers that be determine is optimal; and even more, providing for the extension of physical life without regard to the human soul and spirit. At times, we benefit from these earthly innovations, while at other times moral dilemmas are the consequence. Where do we draw the line between acceptable and unacceptable technological advances in our world, the choices we make, how humanity views the universe, and whether or not we must embrace these ideas too?

Instead of allowing world leaders or secular intellectuals to draw the line for us, as Christians we should turn to God's Word. When "rightly divided", God's Word offers a much better set of guidelines. Sure, it may mean that there is a restriction or two we might question; however, avoiding the pitfalls of an untethered approach to life and its ultimate meaning is more than worth it.

In this chapter, I want to look at several dramatic innovations being researched today that challenge us to consider where they are headed and if we should "go there". My hope is you will ask yourself, "Is this technology crossing the line? Would God approve? If not, what can we do about it?" As of now, I believe the best thing we can do is to be informed and prepared to make choices

Figure 61 - CHERUBIM CHARIOTS BY JOSH PECK

in light we what we can discern while being mindful that technology always begets the possibility of new "evils" as well as new "goods".

To begin, let's understand the origins of technology. Quoting from chapter 2 of my book, *Cherubim Chariots*:

> There is a common tradition that teaches the Serpent of the Garden of Eden was the one responsible for the development of ancient and modern technology. This view stems from certain verses in the third chapter of the book of Genesis:
>
>> [17] And unto Adam he said, Because thou hast hearkened unto the voice of thy wife, and hast eaten of the tree, of which I commanded thee, saying, Thou shalt not eat of it: cursed is the ground for thy sake; in sorrow shalt thou eat of it all the days of thy life;
>>
>> [18] Thorns also and thistles shall it bring forth to thee; and thou shalt eat the herb of the field;
>>
>> [19] In the sweat of thy face shalt thou eat bread, till thou return unto the ground; for out of it wast thou taken: for dust thou art, and unto dust shalt thou return. (Genesis 3:17-19)
>>
>> [23] Therefore the LORD God sent him forth from the Garden of Eden, to till the ground from whence he was taken. (Genesis 3:23)
>
> The common interpretation of verse 19, specifically where it says "in the sweat of thy face shalt thou eat bread," is that Adam would now have to work to prepare his own food. No longer would it be readily available as it was in the Garden of Eden; if Adam wanted bread, he would now have to make it himself. Also, verse 23 states that they had to till the ground. Since tilling ground requires tools and innovation, and since the Serpent was the one who tempted them into this position, it is commonly believed that the development of technology resulted from the influence of the Serpent.[1]

PARALLEL AND EXTRA DIMENSION DETECTION

Many see the LHC at CERN as a giant inter-dimensional portal. Anthony Patch, my colleague in writing this book, speaks specifically to issues related to CERN. However, permit me to provide a simple overview and relate it to our topic. Back in 2010, Reuters reported:

---

[1] *Cherubim Chariots* by Josh Peck, available at www.sharpeningreport.com

> Guido Tonelli, spokesman for one of the CERN specialist teams monitoring operations in the vast, subterranean LHC, said probing for extra dimensions – besides length, breadth, height and time – would become easier as the energy of the proton collisions in it is increased in 2011.[2]

One of the many goals at CERN is the discovery of extra spatial and/or temporal dimensions. The CERN press office states:

> As far as we know, we live in four dimensions, three of space and one of time. But experimentalists at the Large Hadron Collider are looking for evidence that the universe contains more than that. The existence of extra dimensions could explain some puzzling properties of the universe.[3]

Another possible discovery concerns the possibility of parallel universes. Experiments planned with the LHC may lead to this colossal discovery. Though physicists involved with or having special knowledge pertaining to CERN will use the term "parallel universes", they admit to talking only about realities in higher dimensions within "this universe". One author of the study went on record to say:

> Normally, when people think of the multiverse, they think of the many-worlds interpretation of quantum mechanics, where every possibility is actualized. This cannot be tested and so it is philosophy and not science. This is not what we mean by parallel universes. What we mean is real universes in extra dimensions...as gravity can flow out of our universe into the extra dimensions, such a model can be tested by the detection of mini black holes at the LHC. We have calculated the energy at which we expect to detect these mini black holes in gravity's rainbow [a new theory]. If we do detect mini black holes at this energy, then we will know that both gravity's rainbow and extra dimensions are correct.[4]

As I asserted in my book Quantum Creation, whatever scientific discoveries are made, we should assimilate such breakthroughs to strengthen our faith, not shake it. This is why knowing how to separate observation from interpretation is so important today. We must learn

---

[2] See http://www.reuters.com/article/2010/11/15/us-science-cern-idUSTRE6AE3QU20101115

[3] See http://press.web.cern.ch/backgrounders/extra-dimensions

[4] See http://www.ibtimes.co.uk/mini-black-holes-large-hadron-collider-could-prove-existence-parallel-universes-1493157

how to adopt such possible advances into a biblical worldview that glorifies God instead of increasing doubt about His existence. Even in the most extreme cases, such as the discovery of "parallel universes" (regardless of how they want to define it), we can still transform the threat of upsetting our belief system to reinforce what theists have believed for thousands of years. This is not meant to be a mind game or a semantic trick. A spiritual realm exists and it can be reconciled to extra-dimensional existence. We only need examine the meaning of what scientists contend they believe to be true about the "multiverse" and search within the scripture to reconsider our descriptions of the supernatural world we find there; then place it in this new context science affords us.

In other words, terms like "parallel universes" and "extra dimensions" do not need to frighten us. We have the ability to understand these things without needing a master's degree in Physics or intensive, lifelong study in Astronomy. Furthermore, even if they were able to discover parallel universes (such as is described in the many-worlds interpretation), it would not negate anything biblically, but would instead shout even louder to the glory of God's creation. To illustrate this further, consider this excerpt from Quantum Creation:

On the surface, there are some theological problems with this theory that make most Christians dismiss it as a ridiculous fantasy. Questions come up, for example, like "if there are infinite versions of myself out there, wouldn't that mean there are some universes where I have never and will never accept Jesus as my Savior?" and "If every time I make a choice, a parallel version of myself is making the opposite choice, doesn't that negate free will?" as well as many others.

I believe these are very valid questions that deserve an answer, but I do not believe these questions alone are grounds to dismiss the many worlds theory completely. I will state up front that I do not believe there are multiple versions of ourselves out there and I will explain why briefly a bit later. First, let's imagine hypothetically that one day in the future it is proven there are, in fact, parallel universes and they do contain multiple versions of ourselves. Should this be enough to shake our faith? Does this type of conclusion negate God completely? Of course not – it would only prove that God's creation is far more complex than we originally thought. To briefly answer the two questions from the beginning of the previous paragraph: yes, it would probably mean there are other versions of ourselves that are not saved, but this would be on a purely biological level. We have to remember these parallel universes would still be constrained to three dimensions of space. The other "you" might look and sound like you, but they would still have a separate spirit and soul that is unshared. This also answers the second question of free will. It does not negate your personal free will because you still have the choice to do whatever you want, even if the other "you" chooses something different. They would be doing so completely independent of you, and you completely independent of them. This type of theory is nothing that should make us question our faith in God or knowledge of His creation.

QUANTUM COMPUTING

The fact is, whether their intentions are malevolent or benevolent, human beings, no matter how smart, simply do not have enough information to calculate accurately the consequences of their experiments. It is the same with another emerging technology known as "quantum computing". Quantum computing relies on obscure words and phrases like superposition and quantum entanglement to operate.[5]

Let's take a closer look at what scientists mean by these terms. From my book, *Quantum Creation,* I explained these once far-fetched notions that are now accepted science:

> Another interesting thing about photons is their ability to seemingly communicate with each other by means of something at an instant. Einstein called this "spooky action at a distance", yet we would recognize it today as "quantum entanglement". Some-how, by means that are not yet understood, photons seem to be able to communicate vast distances

---

[5] See http://en.wikipedia.org/wiki/Quantum_computing. **For additional reading,** see the following article published by *Wired Magazine*. Retrieved from http://www.dwavesys.com/media-coverage/wired-age-quantum-computing-has-almost-arrived on 9-12-2016.

at faster-than-light speeds. This, of course, would seem to be theoretically impossible. No wonder Einstein was not a fan of this idea of quantum entanglement.

However, by various means, it has shown to be true. Regardless of how far away they are from each other, photons seem to be able to communicate with each other instantaneously. To this day, how exactly this works is a complete mystery. [6]

Again, we see an example of humanity attempting to come to terms (literally) with something it does not have a framework to comprehend. From my vantage point, the only way a notion like quantum entanglement could be possible would be if it were to operate in a spatial di-

Figure 62 – D:WAVE– MAKER OF QUANTUM COMPUTERS
Any reason why we should worry that it looks like a black cube?

---

[6] See *Quantum Creation* by Josh Peck, available at www.sharpeningreport.com.

mension *that transcends the structure of the universe as we conceptualize it today*. This suggests that the Cosmos remains a mystery at its most fundamental level. *This should make all cosmologists humble.*

For example, imagine a two-dimensional universe called Flatland.[7] If I, as a three-dimensional being (speaking only of conventional spatial dimensions), were to stick my finger into Flatland, a Flatlander would only see a line appear before him, and through inspection, would be able to determine it is a circle. However, this is just a single, two-dimensional slice of my actual finger. Now imagine if I were to stick two fingers into Flatland. The Flatlander would have no way of knowing the two circles he now sees are parts of the same higher-dimensional object. He would assume they are similar, possibly even the same sort, yet still separate things.

Figure 63 - *QUANTUM CREATION* BY JOSH PECK

We can use the Flatland example to help visualize higher dimension, as I explain in depth in *Quantum Creation*. If we see two three-dimensional spheres before us, we might assume they are similar, yet distinct. However, we might be mistaken, because these spheres could just be three-dimensional slices of an object made of four (or more) spatial dimensions. This can help explain something as strange as quantum entanglement. If the entangled particles were actually just three-dimensional slices of a single higher-dimensional object, we would not have two separate particles exchanging information; we

---

[7] See Flatland: A Romance of Many Dimensions by Edwin A. Abbott, 1884.

would have two slices of the same object sharing information simultaneously.

If this view has any validity (and I admit, that is a very big "if"), then we really can't grasp what we are encountering – it remains elusive. We cannot comprehend the entire higher-dimensional structure that makes such quantum entanglement possible. Furthermore, we have little to no idea what the consequences are in trying to use entangled particles for the specific purposes of electronic computations (i.e., computing).[8] In fact, recently it has even been admitted that the basic building blocks of quantum computing are ominously "highly unstable", and yet scientists and engineers continue this pursuit without fear.[9] Humanity is probing (I should say "punching holes") into higher dimensions, without respect for the consequences of what may come back through the holes we punch!

QUANTUM TELEPORTATION

We find what is a similar cavalier approach in the research related to particle teleportation, sometimes called "quantum teleportation". One definition, according to Wikipedia, is as follows:

> Quantum teleportation is a process by which quantum information (e.g. the exact state of an atom or photon) can be transmitted (exactly, in principle) from one location to another, with the help of classical communication and previously shared quantum entanglement between the sending and receiving location.[10]

---

[8] Editor's note: As this book was being readied for publication, both the Chinese and the Russians celebrated the fact they had successfully placed satellites in orbit creating a communication network employing *quantum communications*. The satellite in the sky can talk to a communication station on the ground through entangled particles to create and share cyphers, i.e., encrypted keys, providing fool-proof encryption. See http://fortune.com/2016/08/16/china-quantum-satellite-launch/. If messages are intercepted, they are destroyed (utilizing what might be called "the rule of non-observation" – one can't peek at quantum particles without changing the particles observed).

[9] See http://www.nytimes.com/2015/03/05/science/quantum-computing-nature-google-uc-santa-barbara.html.

[10] See http://en.wikipedia.org/wiki/Quantum_teleportation.

Here we have an example of employing quantum entanglement. The most important thing to realize about quantum teleportation: this interaction is no longer confined to the realm of science fiction – it exists today. In 2014, CNET reported:

> Physicists at the Kavli Institute of Nanoscience, part of the Delft University of Technology in the Netherlands, report that they sent quantum data concerning the spin state of an electron to another electron about 10 feet away. Quantum teleportation has been recorded in the past, but the results in this study have an unprecedented replication rate of 100 percent at the current distance, the team said.
>
> Thanks to the strange properties of entanglement, this allows for that data – only quantum data, not classical information like messages or even simple bits – to be teleported seemingly faster than the speed of light... Proving Einstein wrong about the purview and completeness of quantum mechanics (it) is not just an academic boasting contest. Proving the existence of entanglement and teleportation – and getting experiments to work efficiently, in larger systems and at greater distances – holds the key to translating quantum mechanics to practical applications, like quantum computing. For instance, quantum computers could utilize that speed to unlock a whole new generation of unprecedented computing power. [11]

VACUUM (ZERO-POINT) ENERGY

Another discovery on our horizon, one that could be far more practical for most of us, is in the field of renewable energy sources. Scientists (and many conspiracy theorists) consider the most elegant renewable energy source by the term zero-point energy or, perhaps better labeled, Vacuum Energy.

Wikipedia explains zero-point/vacuum energy with these words:

> Zero-point energy, also called quantum vacuum zero-point energy, is the lowest possible energy that a quantum mechanical physical system may have; it is the energy of its ground state. All quantum mechanical systems undergo fluctuations even in their ground state and have an associated zero-point energy, a consequence of their wave-like nature. [12]

---

[11] See http://www.cnet.com/news/scientists-achieve-reliable-quantum-teleportation-for-the-first-time/.

[12] See http://en.wikipedia.org/wiki/Zero-point_energy.

THE FUTURE QUANTUM WORLD

Upon first hearing, this concept sounds awfully convoluted. Basically, the idea conveys that the smallest amount of energy possible for things at a quantum level ("smallest of the small") is its zero-point energy. Because things at a quantum level behave as waves, its energy increases and decreases. The point at which it decreases the most is its ground state (zero-point energy). Think of this like a bouncing ball. When the ball rests upon the ground, its energy then is at zero. This would be, therefore, the ball's zero-point energy.

Zero-point energy in and of itself is merely an attribute of the quantum world. However, vacuum energy is something that actually could be put to use practically in the future. Wikipedia defines vacuum energy as follows (it is confusing, but bear with me):

> Vacuum energy is the zero-point energy of all the fields in space, which in the Standard Model includes the electromagnetic field, other gauge fields, fermionic fields, and the Higgs field. It is the energy of the vacuum, which in quantum field theory is defined not as empty space but as the ground state of the fields. In cosmology, the vacuum energy is one possible explanation for the cosmological constant. A related term is zero-point field, which is the lowest energy state of a particular field. [13]

A much simpler and more effective definition can be found in the same article at Wikipedia. It reads as follows:

> Vacuum energy exists as the underlying background energy throughout the entire Universe. One contribution to the vacuum energy may be from virtual particles that are thought to be particle pairs that blink into existence and then annihilate in a timespan too short to observe. They are expected to do this everywhere, throughout the Universe. [14]

Physicists and other scientists know little about Vacuum energy, perhaps less than any other theory they study seriously today. I should point out, however, that there was an attempt in the past to predict the zero-point of vacuum energy, however, based on actual measurements of the vacuum energy density, it turns out the prediction was

---

[13] See https://en.wikipedia.org/wiki/Zero-point_energy.
[14] Ibid.

off by over 100 magnitudes. (Perhaps we could compare this to missing by a literal mile the tiniest threading of a needle). Experts call this colossal error the "vacuum catastrophe". It has been described as "the worst theoretical prediction in the history of physics." [15]

Despite this conceptual disaster, scientists continue to study Vacuum energy. And one day their work will likely result in advances in learning and the development of valuable technological applications.

In fact, *Business Insider* reported just last year (2015) that vacuum energy might lead to detecting and employing gravitons (the theoretical particle of gravity) and even the elusive "theory of everything":

> It may be possible to draw energy from a vacuum using gravity, a theoretical physicist says. If researchers succeed in showing that this can happen, it could prove the long-postulated existence of the graviton, the particle of gravity, and perhaps bring scientists one step closer to developing a "theory of everything" that can ex-plain how the universe works from its smallest to largest scales. The new research specifically found that it might be possible to show that gravitons do exist by using superconducting plates to measure a phenomenon with the esoteric name of "the gravitational Casimir effect" ... "The most exciting thing about these results is that they can be tested with current technology," study author James Quach, a theoretical physicist at the University of Tokyo, told *Live Science*. [16]

With so much still uncertain about vacuum energy, it is difficult to pin down exactly what, if anything, can be utilized from its discovery. If the most optimistic theories are correct, vacuum energy could provide the world with an infinite, clean, and incredibly powerful source of energy. The bigger questions are, "What would humanity do with this type of power? How would it change the world?"

The vast amount of energy potentially available is only one half of the entire topic. The other half has to do with the discoveries that could come from the utilization and study of vacuum energy. To be more specific, if scientists can ultimately observe gravitons and prove their

---

[15] Hobson, M., Efstathiou, G., & Lasenby, A. (2006). *General Relativity: An introduction for physicists (Reprinted.)*. Cambridge University Press. p. 187.

[16] See http://www.businessinsider.com/looking-for-gravity-particles-2015-3.

existence, it would increase exponentially their knowledge of how gravity works – not only on a macro scale but on a quantum scale as well. This means they would be one very large step closer to learning how to overcome gravitational effects (think flying saucer). It is not outside the realm of possibility (indeed it is likely) that this type of discovery could push weapons technology to far more advanced levels must faster than we can imagine.

Figure 64 - RAY KURZWEIL: THE SINGULARITY BY 2045?

And I can't resist making a comment on the seemingly "holy grail" in Physics known as the *theory of everything*. If scientists are able to discover how to merge quantum field theory with general relativity, the implications are most unsettling. Such a discovery would be so powerful it very well could lead to Ray Kurzweil's *singularity*, predicted to occur in 2045 (the notion that knowledge and progress in technology grow beyond humanity's ability to maintain control).[17] This eventuality is one of which we should be aware, but precisely how we should prepare for it is anybody's guess (some might argue

---

[17] See http://www.2045.com/.

the only solution is to "go off the grid"). As we trek forward along the path inexorably to the future, we head into strange world indeed.

CONCLUSION

Emerging technologies are being developed with little to no regard for their consequences. Whatever comes our way, good or bad, we are the ones who will suffer the consequences. Proverbs 13:16 says: *"A wise man thinks ahead; a fool doesn't, and even brags about it!"* (*The Living Bible*) With the continued acceleration of scientific understanding, and the frightening possibility of the so-called "singularity" rapidly approaching – that moment where technology goes out of our control – it is the responsibility of each and every one of us to look into these things and prepare ourselves and our children accordingly.

It is inevitable that science will bring forth new discoveries. It will also seek to interpret the meaning and force us to accept the implications of those discoveries in light of its own worldview – a worldview usually antagonistic to that of the Bible. As informed theists (and even more as Christians dedicated to living in light of God's truth), we do not have to agree with the interpretations science offers; neither do we necessarily have to employ every new application or device created from these new technologies. Indeed, much of our challenge in this "brave new world" is maintaining our moorings – making sure we don't break free and become caught in the deadly whirlpool caused by technology's rapid advancement. To switch metaphors, the Bible has been a source of humanity's grounding for over a hundred generations. It is a sure foundation. There is no need to seek higher ground.

Likewise, we are not required to heed the wisdom of the world. And follow its lead wherever it's headed. We should remember that no matter "whatever the world is coming to" – *we don't have to be where it's going when it gets there!* That is what Paul meant when he admonished us, *"And be not conformed to this world: but be ye transformed by the renewing of your mind, that ye may prove what [is] that good, and acceptable, and perfect, will of God."* (Romans 12:2)

ABOUT THE AUTHORS

Anthony Patch is a retired Paramedic, working in Oakland, CA for 27 years in the private sector emergency services. For over 25 years, he has been researching in the fields of physics, cosmology, biology, computer science and theology. Since the publication of his first two novels in a trilogy, *Covert Catastrophe* and *2048: Diamonds in the Rough*, he is now a well-known author and speaker. He is heard regularly on *Caravan to Midnight*, *End Times Matrix News*, *FaceLike TheSun*, *The Hagmann and Hagmann Report*, *The Sharpening Report*, *Truth Frequency Radio* as well as many others. With over 400 hours of interviews, he is heard weekly around the world. Anthony is best known for his extensive, leading-edge and revelatory research focused upon the work being done in Geneva, Switzerland with the Large Hadron Collider (LHC), part of the CERN organization, specifically as these experiments relate to the fulfillment of Bible prophecy.

Josh Peck is an avid researcher of "fringe" topics, videographer at *SkyWatchTV*, creator of *The Sharpening Report (TSR)* recording over 100 programs and acquiring thousands of subscriptions on YouTube (TSR is now hosted by James DeWitt), host of *Into the Multiverse*, and is the author of numerous books, including *Quantum Creation: Does the Supernatural Lurk in the Fourth Dimension?* and *Cherubim Chariots: Exploring the Extradimensional Hypothesis*. Josh has been featured on TV and radio shows, including *SkywatchTV* and *The Hagmann and Hagmann Report*. Josh has spoken at numerous conferences. And he is a family man, married to Christina Peck and has three children: Jaklynn, Nathan, and Adam. Josh works in full time ministry, dedicating his life to *SkyWatchTV*, *Into the Multiverse*, writing books, and providing for his family.

ABOUT THE AUTHORS

Gonzo Shimura is a Christian documentary filmmaker, host of the Canary Cry Radio podcast, and runs the YouTube channel, *FaceLikeTheSun (Currently over 19 million channel views)*. His independent documentary film *'AGE OF DECEIT: Fallen Angels and the New World Order' (2011)* became a YouTube sensation in the alternative Christian community with multiple millions of views and shares. His second film, *AGE OF DECEIT 2: Alchemy and the Rise of the Beast Image (2014)* also has over 1 million views on YouTube. He is the winner of the 37th Annual Telly Award for his work as Director and Editor of the *SkyWatchTV* documentary film *INHUMAN: The Final Phase of Man is Here*. Gonzo is in full time digital ministry preaching the Gospel message of Jesus Christ in light of "fringe" topics like the so-called secret space program, UFOs, aliens, angels, supernatural/paranormal phenomena, the New World Order, transhumanism, and the coming "technological singularity". Gonzo and his wife Erin live in Orange County, CA with daughter Kira (*and another girl on the way as of this writing*).

S. Douglas Woodward (B.A., Th.M.) is a publisher, author, speaker, and researcher on the topics of alternate history, the apocalypse and culture, as well as biblical eschatology with an emphasis upon America's place in Bible prophecy. He has studied, researched, and taught over 40 years on these subjects. He has written twelve books including co-authoring the bestseller, *The Final Babylon* (2013) and author of *Lying Wonders of the Red Planet* (2014) and *Uncommon Sense* (2014). His latest book is entitled, *The Next Great War in the Middle East* (2016). He also contributed to three titles for Defender Publishing including Pandemonium's Engine (2011) and *Blood on the Altar* (2014). Doug occasionally hosts *Prophecy in the News* and appears there frequently as a guest. Doug has served as an Elder in both the Reformed and Presbyterian churches. Currently, Doug serves as an adjunct professor for a major university teaching on the topic of entrepreneurship. Doug has over 40 years' experience in the business world, serving as an executive for Microsoft, Oracle, Honeywell Bull, and as a Partner at Ernst & Young. Doug is married to wife Donna for 41 years, lives in Oklahoma City, two married children, and two wonderful grandsons.